MASTERING SCIENTIFIC WRITING

SECRETS FOR SUCCESS IN THE AGRICULTURAL, BIOLOGICAL, AND HEALTH SCIENCES

WITH HINTS FOR WRITERS IN ALL FIELDS

Robert F. Kahrs, DVM, PhD

Copyright © 2008 by Robert F. Kahrs

To use portions of this work, please contact:

Bob Kahrs

PO Box 840039, St. Augustine, FL 32080-0039

(904) 471-6735

Email: bobkahrs@aol.com

Library of Congress Cataloging-in-Publication Data

Kahrs, Robert, F.

Mastering Scientific Writing

ISBN 0-7414-4695-2

1. Writing—Scientific or Technical. 2. Authorship. 3. English Language—Writing

Published by:

INFI(∞)ITY
PUBLISHING.COM

1094 New DeHaven Street, Suite 100
West Conshohocken, PA 19428-2713
Info@buybooksontheweb.com
www.buybooksontheweb.com
Toll-free (877) BUY BOOK
Local Phone (610) 941-9999
Fax (610) 941-9959

Printed in the United States of America

Printed on Recycled Paper

Published April 2008

CONTENTS

PREFACE

Professionals are judged by their writing. Despite their potential to produce excellent documents, some hinder their careers by postponing writing.

Mastering Scientific Writing lays out a stepwise plan using personal preferences to overcome anxiety and unleash writing potential. It challenges tradition, emphasizes positive approaches, stresses reader comfort, and helps make writing an enjoyable experience. It is based on the experiences of teachers, professors, colleagues, and the authors whose works are cited. These people encouraged me to pursue writing and its professional rewards.

I am grateful to my granddaughter Kierston Kahrs for designing the cover and to editors Cindy Kahrs (my daughter), Beth Mansbridge, and my wife Evelyn.

Most of all, I thank Evelyn for her patience, love, and support in my lifelong struggle to master writing and to eventually produce this book.

INTRODUCTION

Writing excellence leads to personal satisfaction and professional success. Many people think writing ability is a gift. While some people take to writing more readily than others, it is actually a developed skill that improves with practice. Some writers accommodate readers with easily read documents. Others write well but neglect crucial polishing steps. Some bright or verbally articulate individuals can find writing a challenge.

Many writings, such as scientific books, journals, and regulations, are viewed with skepticism because of their ponderous language. These documents differ from all others because they must be both factual and easy to read.

This manual presents positive approaches and helpful hints for writers in all fields. It answers questions and offers flexible suggestions. The first nine chapters outline a stepwise pathway to successful writing. They can be read and implemented in order. The three reference chapters (10, 11, and 12) discuss spelling, punctuation, and grammar. You should read their opening pages, but they are mostly for reference.

The Value of Effective Writing

Effective writing enhances job satisfaction. It is essential in academic, corporate, and governmental careers and is readily achievable. Professionals must choose between writing excellence and a lifetime of struggle. Avoiding writing is not an option. Publish-or-perish systems banish talented speakers, effective teachers, outstanding clinicians, and superb researchers who succeed at everything but writing.

Writing Challenges

People for whom learning comes easily often are not accustomed to the repetition needed to produce excellent manuscripts. Other people—who are used to struggling—may find writing easier. Good speakers can also be challenged by writing. In speaking, statements are emphasized with voice shifts and body language. In writing, emphasis comes from word selection, word placement, and punctuation.

Many excellent thinkers and speakers avoid writing because they doubt their writing ability or hate to write. These aversions can result from self-imposed obstacles like distaste for grammar, belief that there is only *one* way to write, or unrealistic views of the effort required.

Writing excellence is achieved with clear goals, positive attitudes, and practice with everyday communications. Everyone can become a good writer. Computers help by expediting writing, revision, and editing, and by monitoring and correcting spelling, punctuation, and grammar.

The Road to Writing Excellence

The components of writing excellence are brevity, clarity, consistency, precision, accuracy, flow, and style. These ingredients lead to fine documents by blending words into concise sentences and paragraphs. The road to writing excellence improves as personal strategies emerge. Initially the path resembles an obstacle course, but barriers are soon overcome and writing blocks turn into writer's clout. Writing success is achieved by typing rough drafts, conducting multiple revisions, using layaway periods, editing carefully, and proofreading intensely.

To begin your journey to writing excellence, first assess your understanding of the fundamentals of good writing with the self-administered scorecard in chapter 13. Then become familiar with the components of writing excellence described in chapter 1, and you'll be on your way.

CHAPTER 1

THE COMPONENTS OF WRITING EXCELLENCE

Introduction

The components of writing excellence are brevity, clarity, consistency, precision, accuracy, flow, and style.

Brevity

Brevity (briefness) requires statements that are short and to the point. It is becoming crucial as readers increasingly choose short papers and small books. Brevity is best achieved with simple words. The enemies of brevity—wordiness, meaningless modifiers, and redundancy—impede readability and comprehension. The road to brevity begins with revising drafts by deleting cumbersome and nonessential words, as described in chapter 5.

Clarity

Clarity is the art of making clear statements. It depends on word choice and placement and is often missing in drafts. Clarity improves with revision and editing as writers analyze sentences for vagueness, ambiguity, or confusion. Strategies for achieving clarity are detailed in chapters 5 and 6. Clarity partners with brevity to create concise documents.

Consistency

Consistency implies agreement and compatibility among parts of documents. It is achieved by assuring agreement between citations and reference lists and by standardizing headings, formats, and fonts. It is targeted during revision and editing.

Parallelism or parallel structure is one form of consistency. It lets readers focus on content by making similar portions of documents uniform so headings, lists, and long sentences are written alike. Parallelism expresses ideas of equal rank in similar ways. If a list of words separated by commas begins with a plural noun, the entire list should contain plural nouns. If the first

1

heading is in bold caps, all same-level headings should be similar. Parallelism is detailed in chapters 5 and 12.

Precision

Precision is the use of scrupulous detail, certainty, and definiteness. Precise writing is clear, finely honed expression that is free of vagueness (Meyer and Meyer 1993). It uses specific names and measurements rather than general terms. Revision for precision is detailed in chapter 5. Note that false or inaccurate information can be written in precise terms, so each sentence must be evaluated for both precision and accuracy.

Accuracy

Accuracy is the correctness of information, processes, names, addresses, dates, places, and compounds. It requires correct spelling and errorless arithmetic. It is a major goal of revision and editing.

Fear of being wrong can drive competent individuals from writing, but accuracy is easily achieved. Its ingredients (the facts) are available in refereed journals and textbooks. Arithmetic and spelling are easily checked. The confidence of knowing your material is correct will magnify your writing power.

Flow

Flow is smooth, liquid-like movement from word to word, sentence to sentence, and paragraph to paragraph. It helps readers concentrate on logically placed words and thoughts. Smooth flow is important in writing because it minimizes interruptions, redundancies, and excess details. Good flow is achieved by multiple revisions, as described in chapter 5.

Style

There are publishers' styles and authors' styles. A publisher's style involves the required format and organization of documents. Styles for technical journals vary and are usually spelled out in publications. A text on publishers' style is *Scientific Style and Format* (Council of Scientific Editors 2006).

An author's style involves a unique manner of expression that results from personal blends of brevity, clarity, consistency, and flow. It reflects an author's personality, vocabulary, and concern for reader comfort. You can create your own style as long as it doesn't misuse accepted American English (Steinmann and Keller 1999). If you are British, your style should follow accepted English usage (Austin 1999).

Writing styles emulate speech patterns to some extent. However, writing and speaking differ markedly. Many verbal expressions are inappropriate in writing, and good speakers can have difficulty writing. In speech, emphasis is conveyed by gestures, body language, volume modification, changes in tone, and audible pauses. In writing, these effects must be achieved by word selection, word placement, sentence structure, and punctuation. Writers should remember that reading differs markedly from listening. Writers must organize words more carefully than speakers.

Writing styles improve with experience. There are few guidelines for effective writing style. Readers recognize documents that are easily understood. They will point out gifted writers without realizing that writing style results from constant practice and repeated editing.

The classic text on writing style is *The Elements of Style* (Strunk and White 2000). It indicates that writing style results from:

- doing what comes naturally;
- having a specific goal for each document;
- focusing on reader comfort;
- avoiding exaggeration and pretentious language;
- maintaining consistency in lists, headings, and sentences; and
- revising incessantly for clarity and accuracy.

When these tactics are applied, your unique personal style will emerge.

Applying the Components of Writing Excellence

Writing excellence requires knowledge of its components, dedication, and constant practice while drafting (chapter 4), revising (chapter 5), editing (chapter 6), and proofreading (chapter 7). You can practice these skills with routine correspondence and with documents intended for publication. The essentials of common documents are described in chapters 8 and 9.

CHAPTER 2

THE JOURNEY TO WRITING EXCELLENCE

Introduction

Well-written documents suggest that authors had appropriate ingredients, used them in proper proportions, and skillfully blended them to satisfy readers, editors, and publishers. Like a luscious cake, good writing arouses the appetite. It is fun to sample, stimulates a desire for more, and leaves you wishing you had the skills and recipes to produce similar products. You have the skills and this book has the recipes.

The journey to writing excellence involves preparing, drafting, revising, editing, proofreading, and submitting. Preparation and drafting are constructing steps. Revision, editing, and proofreading are polishing steps.

Make writing excellence a personal goal and take action to see that it happens. This requires mental and physical preparation, positive attitudes, and clear strategies. Physical preparation involves space and materials. Mental preparation requires committing time and energy, and realizing that the more we write, the easier it gets. Mental preparation begins with goal setting.

Goal Setting

Clear goals are essential. We seek personal goals of family harmony, financial security, and professional growth. Professional goals include: steady employment; a reputation for scientific acumen, professional credibility, and integrity; invitations to speak and apply for jobs; and assignments to represent our organizations on policy issues. Writing excellence contributes to all these and is readily accomplishable. Goal setting works best if aspirations are recorded. Identify and list your professional goals in a private notebook. Include writing excellence as a goal and list titles of documents you want to write.

Recall Ebenezer Scrooge's visits from ghosts of the Christmases past, present, and future. Imagine a visit from the ghost of your writing past and recall memories that haunt you. If you remember enjoyable writing experiences, you are fortunate. More likely there will be visions of red-marked papers, demanding teachers, and frustrations. No wonder people detest writing. Discuss your current writing attitudes with the ghost of your writing present. Then have a chat with the ghost of your writing future. Will this conversation project writer's clout or writing blocks? Both are strenuous but writer's clout leads to professional advancement while writing blocks lead to stagnation. Your imaginary advisor will tell you to choose writer's clout. Set a date for an annual visit with the ghost of your writing future, record it in your notebook, and keep that appointment each year.

Jot down achievable goals—including successful writing. Then start down the road to writing excellence. The first step is mental preparation.

Mental Preparation

Writing is the summit of a hierarchy consisting of hearing, reading, experiencing, speaking, and writing. Writing consolidates facts, concepts, and verbal skills. It provides personal satisfaction and professional success. Mental preparation for writing success requires recognition of your strengths, positive approaches, can-do attitudes, and daily efforts.

Identify Your Strengths

List the areas in which you have unique knowledge and experience in your field. Choose topics about which you are knowledgeable, concerned, and passionate. Add ambition and you have all the prerequisites for writing excellence.

Positive Approaches

Most people have been schooled in negativity and some believe they can't write. Positive approaches add unique dimensions to writing. Forge ahead knowing you have what it takes. Forget the teachers who barked about *the only* way to write, and try multiple approaches until you find comfortable strategies. During revision, convert negatives to positives. This will overflow to work and family.

Can-Do Attitudes

Convince yourself that you can write well. Dismiss admonitions about what shouldn't be done. Forget the right-wrong scenario and seek good-better-best options. Believe you can write and you will succeed. It won't happen overnight, but each step will produce improvement. Like driving golf balls, the secret is to get lined up, focus on the task, execute necessary actions, and follow through. Every shot won't land on the green and every sentence won't be perfect, but unlike golf, words can be rearranged without penalties.

Schedule a Time for Writing

Few people realize how much time writing, revision, and editing take. Unrealistic perceptions about speed-writing can trigger writing blocks. Answer these questions: Are you sharpest in the morning or at night? Do you have more energy before or after eating? Can you concentrate for several hours? Then dedicate high-energy time for writing every day. Plan a daily hour for undisturbed writing. If you choose morning, get up an hour earlier and write in peace while others sleep. If you are a night person, stay up an hour later.

Some people settle on weird schedules. Many go to bed exhausted and sleep until they wake up. Then instead of tossing and turning, they sneak into a writing hideaway and write till exhausted before returning to bed. Another option is to quit going out for lunch. Instead, bring a sandwich and head for a hideout. While colleagues are eating and gossiping, you can be writing, learning, and advancing professionally.

Have you assigned a date yet for an annual visit with the ghost of your writing future? On the appointed day each year, brag about the writing secrets you've learned and the skills you've mastered. Time with this phantom counselor will help identify good habits and plot the demise of bad ones. It can coincide with completing the scorecard in chapter 13.

Spousal Commitment

Spouses and significant others must support your decision to dedicate time to writing. Talk it over and indicate what it means and when and where you will write for an undisturbed hour five days a week. Identify activities you will abandon to provide this time. Start with TV commercials. Tape TV programs and fast-forward through commercials when watching them. Examine your reading and relaxation habits and redirect time to writing. A daily writing hour won't detract from other obligations.

Join a Writers' Support Group

Most communities have writers' support groups. They present programs and critique members' work. Writing groups aid in identifying books, websites, print-on-demand operations, and methods for approaching publishers. They usually have spelling or grammar experts from whom we can learn a lot once we get over being defensive. Writers' groups can add new dimensions.

Writer's Clout or Writing Blocks

Most obstacles to writing are mental. They consist of under-estimating the required effort, unwillingness to commit time, and succumbing to writing blocks. Our brains handle around 60,000 thoughts daily. Most of them never reach our conscious mind. Each conscious thought triggers others and can drag sequestered ideas from the subconscious. One secret is to generate a thought, type it, and then follow your instincts without looking back until you have recorded things you never realized were there.

Writing blocks are inabilities to get going. They have compli-cated causes and effects. They are not a single affliction but comprise multiple influences that make writers get stuck. They often affect bright people (Hjortshoj 2001) and may occur in short, easily overcome episodes. Writing blocks can result from fear of failure. They promote procrastination and deprive their victims of ability to concentrate. They prevent verbally articulate, highly intelligent, and proud individuals from putting their thoughts on paper. Writing blocks keep true writing skills from emerging and reduce professional confidence. They can be characterized by self-doubt, fear of being wrong, procrastination, perfectionism, uncertainty, or nervous anxiety. Anxiety about writing can actually be a prelude to success. Athletes and speakers admit their performances are best if accompanied by an adrenaline rush. Writing experts and psychologists are indecisive about writing blocks (Hjortshoj 2001) and avoid identifying their place among personality traits, behavioral defects, or other disorders.

The victims of writing blocks may stare blankly but write nothing. Sometimes they write and delete continually. They may get bogged down in revision or get depressed over helpful suggestions. They decide this is not a good time to write, but the scenario continues until they conclude they weren't gifted with writing ability. Some victims undertake any task that arises as an excuse to postpone writing.

Writing blocks are easily converted to writer's clout (see chapter 3). Writer's clout is power generated by those who develop writing skills and use them productively. This clout produces leverage at work. It opens new doors. Excellent writers are rare in most disciplines, and those who write well and willingly, contributing their talent to support their organizations, swing incredible weight. When we choose writer's clout we:

- develop a relentless ambition to release our unfulfilled potential;
- unleash hidden energy and become driven taskmasters;
- change our expectations about the time that writing requires;
- establish a writing time and place;
- draft wildly without interruption; and
- revise documents repeatedly and edit them meticulously.

When these challenges are addressed, suppressed talents blossom and new confidence emerges. When we are mentally prepared, physical preparation is easy.

Physical Preparation

Physical preparation requires a private writing site. Athletes work out in weight rooms, artists spend hours at the easel, and musicians practice incessantly. Writers—you guessed it—must practice. This requires a well-equipped studio. Physical preparation also requires body and brain conditioning that arises from adequate sleep, regular exercise, and a sound diet.

Writing Hideaways

Find a distraction-free writing hideaway. Then schedule a daily writing hour and discipline yourself to this schedule. If you choose to write at work, leave your office during your writing hour unless you come in early enough or stay late enough to work without interruption. Seek a library or empty room to work in. You'll probably want a hideaway at home too. You'll also need a writing briefcase stuffed with project folders and notebooks.

Reference Materials

Essential reference materials include a dictionary and a thesaurus, both less than five years old, and books and journals in your area of expertise. Bargain shelves at bookstores contain books on writing and grammar. *Research Papers for Dummies* (Woods 2002) and *Good Grammar Made Easy* (Steinmann and Keller 1995) are good starters. Two valuable references are *The Handbook of Technical Writing* (Alred et al. 2003) and *The Gregg Reference Manual* (Sabin 2001). Writing books offer differing rules and approaches. Don't get excited; many books in other disciplines do the same thing.

Computer Utilization

Earlier generations wrote in longhand or on typewriters. Computers have changed that. They permit professionals to develop a reservoir of verbiage that can be constantly expanded, updated, and reused. Everything you have ever written—if saved in identifiable files and folders—is at your fingertips. If you cannot already find things instantly, clarify your file identification, storage, and backup strategies.

Three-hole-punched hard copies of major documents with their hard drive and backup disc locations recorded at the end will be useful for years, making it easier for you to sort through mountains of text generated by computers that force readers to choose documents by their titles and opening paragraphs. Computers expedite revision and editing by monitoring spelling and grammar.

Spell-Checker

Spell-check programs, sometimes called spell-checkers (Pfaffenberger 2003) are superb. Some typists leave them on as they work to identify typos and spacing errors as they go along. Avoid over reliance on spell-checker programs because they are subject to human error and lack specialized terms (see chapter 9).

Grammar Checker

Grammar checker programs usually run concurrently with spell-checkers. Like spell-checker, they must be used cautiously and require careful oversight by authors. They are good at detecting misplaced, absent, or duplicated periods; run-on or incomplete sentences (fragments); and the passive voice. When grammar checkers underline passages, it means something is unclear. Even if you don't understand what they are saying, it's a good idea to fiddle with the sentence until the underline disappears, or seek an explanation by clicking the spell-checker icon on the toolbar, or by clicking Tools→Spelling and Grammar in Microsoft Word. Spelling and grammar checkers are constantly being improved.

Speech Recognition Programs

If your verbal communication skills exceed your writing ability or if you type slowly, you may want to consider a speech recognition program for your computer. After installation, setup, and training exercises, these devices adapt to your voice and direct the computer to type as you speak. The resulting text—like all drafts—requires heavy editing. This is a great way to initiate manuscripts. You can transform speeches into rough drafts by taping them and playing the recordings into your speech recognition program.

PowerPoint

Microsoft PowerPoint produces automated presentations that integrate written material with photos, artwork, and recordings for slide projection. PowerPoint permits printing handouts with four to five slides aligned vertically on the left of the page and lines for notes on the right. These can be given to audiences or used as ready-made topic headings to expedite transitions from speaking to writing.

Computer Strategies

Personal writing and computer skills grow simultaneously. Gradually you'll develop shortcuts and customized templates for letterheads, memos, trip reports, and emails. Whenever you find a procedure repetitious, take time to explore shortcuts or develop easily accessible outlines.

Develop Flexible Personal Strategies

The procedures described in books about writing may be awkward. Feel free to proceed in whatever order you think of things to write. Let your preferences lead to methods that work and lead to a comfortable personal writing strategy that may include:

- saving everything you write on disk (hard drive) and CD, and later harvesting your work;
- placing asterisks in texts to mark interruptions, placement of moved text, and needed corrections;
- drafting summaries before projects are started;
- keeping a pencil by the keyboard to record sudden thoughts;
- carrying a pocket-sized pad or note cards to record unexpected thoughts; and
- keeping a bedside pencil and pad for ideas that emerge on sleepless nights.

You will find personal strategies to complement this list. Once you've committed to mental and physical preparation and have some goals listed in your notebook, it is time to begin the journey to writing excellence, starting with a simple document and then moving on to a more complex paper. Don't wait, go straight to chapter 3.

CHAPTER 3

INITIAL PROJECTS

Introduction

The journey to writing excellence begins with projects. Write a short document—like a memo or letter— and then a longer one in your usual manner to establish a baseline. If time has passed since you read chapter 2, skim it and the notes you made about your goals. Prepare by choosing subjects, setting goals, and deciding on formats. The computer industry is moving away from floppy disc drives so using compact discs (CDs) for backing up your files will serve you better in the long run.

A First Project

Choose a first project—an email, memo, or letter—and lay out its goals. Jot down the topic and audience, decide what you want readers to know and do, and develop a brief outline. Emails, memos, and letters are best if concluded on one page.

Write Your First Document

Look over the points you want to make and hack out a rough draft. Save files as you go along, incorporating the date in the file name for clarity. Next, undertake three quick revisions. First, revise the drafts so that your points are in logical order. Second, search for and delete unnecessary words. Third, make sure each sentence is accurate.

Next, proofread it line by line and search for proper spelling, capitalization, and punctuation. Then lay it aside. After two days, retrieve it, read it over quickly, make improvements, save it to a floppy disc or CD and send it. Print two hard copies, one for posterity and one to edit a year from now. Label them number one in red ink, three-hole-punch them, and put them in your writing notebook.

Don't worry if you are not comfortable with the sequence just described. There is no single right way. Your preferences will change as you progress.

Now repeat the process with a more complicated task such as a speech abstract, a journal article, a book chapter, or a thesis chapter. After the initial draft is complete, save it to disc, lay it away for a week, and then revise it. Label it number two and save it as before.

Plan and Prepare for Your Next Few Projects

After completing these projects, start planning others. Identify three urgent documents—exclusive of letters or memos. If you are a graduate student, the thesis is number one. Start it now, even if the study is just beginning.

If you are further along in your career, list papers you should have written. For each, write the first title that occurs to you. Find a folder with pockets on its inside covers, or grab a manila folder and staple a used greeting card or piece of thin cardboard to the bottom of its inside covers. These form pockets for reprints, outlines, drafts, discs, and CDs. Label each folder with the title and date. These folders are portable nerve centers. They go in briefcases for pondering during idle moments. No one has to see them. They are your evidence that you are on your way and will soon be drafting.

Pick the most important document and list its main topics from start to finish. Preparation may require a literature search or creation of an outline of an experiment or survey. The extent of your preparation depends on the project. Extensive notes may be required. Researchers and graduate students can begin drafting reports or thesis chapters before the work starts. This process identifies needed information, generates new ideas, and gets things started. Number the hard copies and after a few weeks reread and edit them.

Collecting Stuff

Just because you've identified a future masterpiece doesn't mean you must start drafting. There are prewriting activities. They include creating the title, listing the topics, and gathering ideas. Documents require lots of thought. These thoughts can pop up unexpectedly. Scribble them on cards or pads that you will soon begin carrying. Put the notes in project folders as you reconsider the title, the intended message, the audience, and the outline. Collecting is a continual process.

You'll need a pencil and pad at your bedside to record ideas that emerge at night. It is amazing how fast you'll go to sleep once you've jotted down something that you would have forgotten by morning. These collections will shape your outline and draft. The joy of project folders is that you can brainstorm documents on planes, in lunch rooms, or during TV commercials.

Lists and Bubble Diagrams

You may create a list of topics immediately. Some people create bubble diagrams with the title in the center of the page surrounded by tentative topic headings. They circle them and soon have a mess of bubbles that can be connected with lines indicating their relationships and numbers in the order of placement. When the page gets too messy they redo it in a neater, more logical order and toss the original. Others write topic headings on sticky pads and paste them on a wall or mirror and move them around as new thoughts appear. Try these methods and see which is more stimulating for you.

Outlining

Try writing an outline based on items in your topic list or bubble diagram. Then edit and retype it several times and add new items that surface during the process. Memos, letters, and short documents need an introduction, body, and conclusion. Journal articles require the publisher's format and headings.

Organization

Writings must flow logically. The order can be chronological for histories, topical for lecture notes, or sequential for instructions. An outline helps improve flow and expedites transition from topic to topic. Professional journals have required headings but you can vary information within them. Put the most important information near the beginning.

Draft As If You Know What You Are Doing

Finishing the first draft is crucial (see chapter 4). Sit down at the computer and type furiously using your own writing strategy (Thurman 2003). Forget about brevity, clarity, and consistency. Ignore grammar, spelling, punctuation, capitalization, and sentence structure. Don't stew over logical order or about sticking to your outline. Such concerns distract you from getting something on paper (Cazort 2002). This is no time to be picky. Type something under each heading. If you bog down, get up, walk around, grab a coffee, juice, or soda, and come back with new ideas and a can-do attitude. Avoid conversations during these respites. When you've got something written on each item in your outline, number the document in the upper right-hand corner and each time you print another version transfer this number. Then set the draft aside for a week.

Don't rest during this time. Brood over it but don't look at it. Write thoughts about it on cards or paper scraps and stuff them into the document folder. During this layaway period start on something else. While the draft is laid away, ponder questions like whether you should have started with the summary.

Starting with Summary and Conclusions

Summaries and conclusions are usually written when documents are nearly finished, but some authors like to begin by creating a summary. They use its key words as topic headings. A prematurely written summary is one way of focusing. There is more on the summary-first technique in chapter 4.

Turn on the Writer's Clout

The journey to writing excellence opens a new lifestyle. It replaces old hang-ups, frustrations, and writing blocks with courageous enthusiasm, renewed spirit, and writer's clout (see chapter 2).

Writer's clout combines personal confidence with writing quality and yields successes that will surprise and inspire you. As your clout grows, your self-confidence, speaking skills, scientific knowledge, and influence in the workplace all expand. The crucial adage in pursuit of writer's clout is "well begun is half done." Your initial determination and commitment will set a course to success.

Writer's clout requires access to accumulated material about each document. Keep project folders in your briefcase and work on them continually. Permanently valuable writings can be transferred to a document collection on your hard drive and backed up on floppy discs or CDs. This will save you hours once model memos, letters, and reports are created and available for editing.

Another part of writer's clout is acknowledgment that writing blocks happen to everyone. Be prepared. Consider your writing talent a resurrection fern that can dry and shrivel to apparent lifelessness only to burst forth with new growth after rain. You can provide that infusion by shifting gears, working on something else, getting some rest, and returning with renewed excitement.

Once you come to grips with these issues, you are prepared to embark on the journey to writing excellence. Save numbered documents in your notebook so you can read and edit them next year and see the progress you have made.

First drafts are so important that they require a chapter of their own, so read on to learn more.

CHAPTER 4

INITIAL DRAFTS

Introduction

Drafts begin with document preparation (Meyer and Meyer 1993). It is time to start drafting when your project folder is bulging and you have topic headings. Proceed blindly and recklessly to produce a draft that no one else will see. Freewheeling drafting goes best when you are mentally conditioned and your hideaway is equipped (see chapter 2).

Everyone approaches drafting differently. Some play classical, big-band, or rock music while they write. Others need silence. Still others are stimulated by people and draft best in libraries or lobbies. Getting started is the hardest part. The secret is to get at it, move quickly, and type whatever comes to mind under each heading in your outline.

This form of brainstorming taps into your subconscious and generates new thoughts. Resist temptations to polish until drafts are completed. When we hurry, we write like we talk and drafts sound more like rambling conversations than writings. That's fine because during revision you'll aggressively address the differences between speaking and writing.

Some fiction writers create multiple drafts in an attempt to clarify characters and plots. These often reappear as new stories. In nonfiction writing, one draft is usually sufficient since multiple revisions provide adequate opportunities for improvement.

The First Draft

Most writers dread the first draft and try to get it done as quickly as possible. Some find their hideaway, let the music blare, convince themselves they are great writers, and don't stop till they've finished the entire draft. They glance occasionally at their outline and type freely without looking back. Some stake out an evening and work all night. Some rent a motel room and stay till the job is done. Still others schedule three-hour work sessions and stick to these appointments. After what seems like an eternity, they have pages of gobbledygook. It doesn't matter how it looks. Finishing a draft is a major accomplishment.

Deadlines make you hasten. Set a deadline for beginning and completing each draft. Eventually you'll become good at estimating the time needed. The goal of drafting is to get something written any way you can. If it is a report or paper, you'll want to get it done quickly. You may be drafting a book for years.

Drafting Strategies

After you draft a few papers or reports, you'll find a personal strategy. The secret is to forge ahead and save improvements for later. Most of us start correcting and revising too soon. This makes drafting go on endlessly and creates frustration. Keep drafts to yourself. They have mistakes that won't survive revision. They also contain confidence-building stuff that you didn't know was lodged in your subconscious.

Some writers like to skim completed drafts, read them aloud to identify areas needing work, and scribble notes in the margins. If you do this, go through the entire document before retyping. During drafting, disregard personal pride. Don't worry about clarity, spelling, or punctuation. That comes later. It is important, but not during drafting.

There are many ways to start drafts. Start them before you have searched all the literature and done every experiment. Delaying drafting can dilute your focus. It can be a stalling tactic. Stalling is one component of writing blocks (Hjortshoj 2001). Perfectionists may suffer blocks because they refuse to start drafting until they are completely ready and somehow they're never quite ready. Excess preparation inhibits creativity. Some procrastination can be overcome by drafting a summary before beginning the manuscript.

The Summary-First Technique

The summary should briefly mention main points and lay out conclusions and recommendations. Some experts believe summaries should be written only after a document is finished. Others prefer to draft a summary first to guide content and order, and bring out new ideas.

The summary-first technique involves initially drafting a rough summary and reworking it until it covers essential points and appears to flow logically. Each noun can trigger a subheading. If you choose this technique, throw it together fast. When done, it can be incomplete and messy. No one will see it and you'll feel good. Preliminary summaries may need revision after the document is finished.

Summary of Drafting

The simplest drafting process is to tell yourself you're great, shut out negative voices of your past, and start writing. Inspirations result as soon as words appear. Freewheeling rough drafts will unleash your writing potential and dredge long-suppressed wisdom from your subconscious.

Once the draft is done, save a hard copy to compare with your final version. They will seem like different documents and you'll be proud of your revision skills. You will rarely like a rough draft, so lay it aside and get ready for the next step, repeated revisions.

CHAPTER 5

REVISION

Introduction

Revision is the first—and most complex—step in transforming drafts into documents. Unlike drafting, revision requires focus. It involves rearrangement, rewriting, and correcting. Repeated revisions make drafts understandable and reader friendly (Alred et al. 2003).

Revision Goals

Well-organized and understandable writing requires effective word selection and placement, and appropriate punctuation to mold clear sentences into meaningful paragraphs. Revision is time consuming. It demands concentration. Multiple revisions—with intervening layaway periods—are needed before documents are ready for editing and proofreading.

Multiple Revisions

Revision can be fun but you can be tempted to do too much in a single operation. Individual revisions are usually needed to improve:

- content, organization, and flow;
- choice, placement, and spacing of words;
- brevity, clarity, precision, and accuracy; and
- punctuation, spelling, and grammar.

These goals can be accomplished in any order. Attempting them simultaneously can precipitate revision avoidance syndrome—a cause of writing blocks. This is prevented by breaking the task into steps. In time you'll find your favorite sequence and you can list your personal revision strategy in a writing notebook. Whatever order you choose, see that the document contains the desired content, is well organized, and flows logically.

Post-Revision Layaway

After several revisions it pays to set a document aside for a week. Then rework it and make more improvements. The value of layaway periods will quickly become evident.

Revision for Content, Organization, and Flow

Begin with these questions when revising for content, organization, and flow. Who is the audience? What unforeseen individuals—like bosses, competitors, or media—might see it? Does it clearly state essential points and flow smoothly? With these questions in mind, revise for content, organization, and flow.

Content

Check content to assure that:

- the material is consistent with the title and the text sticks to the subject;
- scientific terms and acronyms are defined when first used;
- controversial or new facts are supported by references; and
- citations agree with reference lists in dates, names, titles, and spelling.

These conditions may require rethinking the title or reorganizing some points.

Organization

Documents need standard sentence organization (subject-verb-object order) and paragraph patterns (introduction, body, and conclusion). These can be varied occasionally. Professional journals specify their required formats and author headings. Book publishers provide author guidelines.

Flow

Logically placed sentences with fluid transitions provide smooth flow. Each point needs a separate paragraph with an introductory sentence, several clarifying sentences, and a closing sentence that sets the stage for the next paragraph. Here are some ways to evaluate and improve flow:

- Read documents aloud, listening for agreement between spoken pauses and punctuation.
- Read silently and mark places where you back up. Ask "What's wrong?" Then rewrite.
- See that information in introductory sentences is explained immediately. If opening subjects are replaced in subsequent sentences—by words like *this*, *that*, or *they*—make sure it is it clear what nouns these words replace.
- Identify and eliminate nonessential words and redundant statements.

Rearrangement to improve flow will expedite evaluation of format and style.

Formats and Styles

Format is document arrangement and appearance. Effective formats require consistency of type size and style and similar patterns for underlines, italics, and bold print (see chapter 11). Similar line spaces, paragraph structures, headings, and indentations also are needed throughout documents. Format consistency is improved by standardizing margins, indentations, headings and subheadings, spelling and capitalization, and by text citations and reference lists. Computers permit many formatting options. Formats can inadvertently change as sentences are manipulated or passages are inserted from other documents.

Consistent styles are challenging in multiple-authored documents. Publishers' styles must be followed but authors have flexibility within required headings. When topics shift, insert subheadings instead of simply beginning new paragraphs.

Choice, Placement, and Spacing of Words

After revising for content, organization, and format take a look at the choice, placement, and spacing of words.

Word Choice

Word choice impacts writing. If you like to sound scientific, academic, or sophisticated—be careful. Remember that simple words permit easy reading, clear understanding, and carefree translation to other languages. Secrets for choosing words are:

- find synonyms for repetitive or pretentious words; and
- use plural subjects when possible because they eliminate the need for limiting adjectives like *a*, *an*, and *the*, and also expedite gender neutrality by using *they* or *them* instead of *he*, *she*, or *he/she*.

Word choice is a key to clarity and brevity, and sets the stage for word placement and word spacing.

Word Placement

Word placement determines ease of reading. During revision, place subjects close to verbs, verbs close to objects, and adjectives and adverbs close to the words they modify. Revising to achieve this proximity improves word order and reduces unneeded words.

Word Spacing

Drafts may contain two or more spaces or missing spaces that are correctable during revision. Some organizations still require two spaces after each period. This convention is shifting and most editors seek a single space after the punctuation mark at the end of sentences. The Microsoft Word grammar checker indicates missing or extra spaces with green underlining. When a green underline shows up, fiddle with the spacing until the underline disappears. You may have to consult the spelling and grammar checkers to sort it out. Improve underlined passages even if the spacing error seems irrelevant because the defect will remain and become evident if the text is manipulated during editing or printing.

Revision for the Active Voice

The active voice should dominate writing. It is more direct, easier to understand, and uses fewer words. In the active voice the subject performs the action, as in *The dog bit the boy* (five words). In the passive voice, the subject receives the action, as in *The boy was bitten by the dog* (seven words).

The passive voice is common in bureaucratic documents. It sidesteps subject-verb-object order, increases the number of words, and sometimes omits subjects. The passive voice is fine when the performers of actions are unknown or when confidentiality is essential. Some say the passive voice preserves research objectivity by focusing on the project rather than the scientists (Hodges et al. 1998). Use of the passive voice often comes from prejudice against using the first person (*I* or *we*) in scientific writings.

Clear writing requires skillful use of the active and passive voices. When you spot the passive voice ask yourself why the individual initiating an action has been reduced to secondary importance or made obscure. If you do not know who performed an action and references are unavailable, you may divert to the passive voice. However, if false modesty or unwillingness to accept responsibility is driving the passive voice, think again. A good way to ameliorate long passive-voice sentences is to identify the performers of actions and make them subjects of shorter sentences. Searching documents for the passive voice is a perfect entree for further revisions.

Revision for Brevity, Clarity, Consistency, Precision, and Accuracy

Brevity, clarity, consistency, precision, and accuracy are cross-pollinating components of writing excellence. They can be addressed in a single revision.

Revision for Brevity

Brevity reduces reading time and page numbers. Its enemies are wordiness and redundancy, which are usually prevalent in drafts and are overcome by eliminating nonessential words and repetitive statements. Reducing wordiness starts by searching documents for meaningless modifiers, but be sure to retain the essential limiting articles *a, an,* or *the* along with the essential connecting words *and* and *or.*

Extraneous words can be eliminated by changing: *as well as* to *and*; *possessing the capacity for* to *can*; *the majority of* to *most*; *have permission for* to *may*; *at this point in time* to *currently*; and *due to the fact that* to *because.*

Meaningless modifiers are common in fiction. While less prominent in technical writing, they can surface in drafts and require attention during revision. Meaningless modifiers like *very, big, high, heavy, fast,* or *significant* suggest comparison but beg the question "Compared to what?" Deleting them or replacing them with specific values improves brevity, clarity, and precision.

After clearing away unnecessary words and nonessential modifiers, start hunting for repetitive word use and redundancies.

Reducing Word Repetition

Raise a red flag when the same noun or adjective appears twice in a sentence or three times in a paragraph. Simple modifiers like *a*, *an*, and *the* as well as introductory words like *in*, *to*, and *for* can be used repeatedly. Repetitive words can be replaced with synonyms. For example, replace:

- nation with country, area, region, territory, or place;
- disease with condition, infection, aliment, syndrome, or affliction;
- agriculture with farming, ranching, husbandry, or agribusiness;
- commodities with products, goods, consignments, things, or materials; and
- environment with ecosystems, surroundings, habitats, or conditions.

Similar substitutions—found in thesauruses—will add variety and simplify reading.

Reducing Redundancy

Redundancies—repetitive words, phrases, or explanations—cause readers to think they have lost their place. When they appear, decide where they fit best and delete them from other locations. Limited redundancy is permissible when identical points are widely separated and used in different contexts. Some redundancy is permissible in summaries, but it is best to alter the wording so it doesn't repeat previous language exactly. During revision, watch for word clusters in which several words say the same thing. When sighted, reduce: *each and every* to *each*; *few in number* to *few*; and *mix together* to *mix*.

Another form of redundancy is the use of multiple words with overlapping meanings when a few will do. Years ago, I wrote the following 26-word sentence: *The laboratory carries out research, experiments, and field investigations on diseases caused by viral and bacterial pathogens of cattle, sheep, goats, and other four-stomached mammals.* An editor rewrote it in seven words as *The laboratory investigates infectious diseases of ruminants.*

Clues to Wordiness

Signs of wordiness are long sentences and paragraphs, redundancy, word strings, and pretentious words.

Replace Word Strings with Single Terms

Drafts often contain multiple words in series. Revisions of word strings can change terms like *histopathologic examination of tissue specimens* to *biopsies,* or *complex and variable food consumption patterns* to *diets.* Word strings of similar terms separated by commas should be examined to delete extra words that convey the same meaning.

Achieving Brevity Using Acronyms and Plural Nouns

Carefully placed acronyms expedite brevity. Spell out and define acronyms when they are first used. To pluralize acronyms or initialisms add the lower case letter *s* without an apostrophe.

Wordiness also results from achieving gender neutrality by a he/she escape route. Avoidance of gender awkwardness can be accomplished by initially using plural nouns (like scientists) and subsequently replacing such nouns with the word *they* instead of *he/she*.

Strategies to Improve Brevity

Strategies for improving brevity include using positive phraseology, eliminating nonessential modifiers, dividing long sentences into smaller ones, and replacing singular subjects with plurals. Try to avoid:

- two or more words when one will do;
- long, pretentious words when short, simple words will do;
- weak words that require modifiers when stronger, more specific words will do;
- nonspecific words like *numerous* or *frequent* when actual numbers are available; and
- the passive voice when the active voice is possible.

These strategies increase both brevity and clarity.

Revision for Clarity

Clarity implies clearness. It is achieved by placing simple words in the best possible order. Clarity improves readability and encourages readers to continue after perusing opening paragraphs. It is achieved by searching documents for vagueness, nonsense, pretentiousness, double-talk, gobbledygook, jargon, ambiguity, and overuse of idioms, metaphors, similes, and the passive voice (Steinmann and Keller 1999).

Reducing Vagueness

Vagueness reflects author uncertainty. Vague statements can be improved by shifting from the passive to the active voice and by reviewing sentences for accuracy and precision. Eliminate double-barreled hedges that use combinations of two words—both indicating uncertainty—such as *may possibly* or *somewhat inconclusive* to indicate uncertainty.

Reducing Nonsense

Nonsense rarely appears in technical writings, but hastily created manuscripts can sound less than thoughtful. If sentences appear to be meaningless, absurd, or frivolous, revise them into thoughtful and accurate statements.

Reducing Pretentiousness

Scientists have a tendency to choose sophisticated terminology. Look over documents for expressions that could appear pretentious or self-promoting. Be particularly alert for words ending in the suffix *ous*. It means having or full of. Whenever possible, replace: *incongruous* with *unsuitable*, *fastidious* with *careful*, and *superfluous* with *extra*. Replace words intended to suggest author intelligence. You can improve clarity by replacing: *cognizant* with *aware*, *terminate* with *end*, and *delineate* with *describe*.

Reducing Double-talk, Gobbledygook, Jargon, Idioms, Clichés, Metaphors, and Euphemisms

Double-talk is artificially clever language that sidesteps, disguises, or evades facts. Gobbledygook, jargon, euphemisms, bureaucratic terms, and inflated language are examples.

Gobbledygook is stuffy, vague, and pretentious writing laden with jargon, buzzwords, legalese, and multiple modifiers. It frequently arises within organizations whose members try to sound scientific or authoritative and develop internal languages which are meaningless to outsiders. It appears in legal documents where intentionally vague language is added to cover all contingencies. It is confusing and hard to understand. The fight against gobbledygook has raged for years (O'Hayre 1966). Gobbledygook and its many forms are common in interoffice correspondence but inexcusable in extramural writings. It is reduced during revising and editing.

Jargon is specialized terminology of trades and professions. It is usually understood by a limited few. It is prevalent among computer experts who adapt established words, such as *access*, *bus*, and *cookie* to transmit special meanings. Computer dictionaries describe computer terms. Some experts say "people who don't understand our language shouldn't be reading our papers." This ignores editors, grant reviewers, and the media. Jargon is inappropriate unless each specialized term is defined when first used.

The figures of speech defined as idioms, clichés, metaphors, and euphemisms are temporarily fashionable usages with unique definitions. They convey subtle but unclear meanings that create confusion for readers and translators. **Idioms** are language subtleties, words, phrases, or sayings from regional dialects whose meanings are unclear to nonnatives. **Clichés** are usually comparisons or descriptions that have been overused and no longer convey the desired impact. **Metaphors** are comparisons between dissimilar things intended to clarify unfamiliar ideas by comparing them to opposite but familiar things. **Euphemisms** are words or phrases used in place of offensive, sacrilegious, or taboo terms. All these figures of speech carry subtle meanings that challenge readers and are difficult to translate to other languages. They are replaceable.

Reducing Ambiguity

Ambiguity creates multiple meanings and mutually exclusive interpretations. It comes from words, phrases, or sentences that are unclear, vague, or equivocal. There are many ambiguous words, such as *awful*, *render*, and *hot*. Multiple-meaning words are confusing to translators and people for whom English is a second language. Care is needed to avoid ambiguous adjectives and adverbs.

Ambiguous statements can arise from positioning modifiers or phrases so it is unclear which words they modify. For example: this sentence, *Investigators found cattle dying in pastures with dehydration and low vitamin A levels,* raises questions. Did the pastures have dehydration or just low vitamin A levels and did the cattle have both dehydration and low vitamin A levels? Careful word placement avoids ambiguous sentences like *The prosecution countered the defense with their best arguments,* which raises the question of whose arguments are being discussed.

Consistency and Parallel Structure

Consistency requires compatibility with respect to punctuation, spelling, headings, fonts, and type size. It also requires that references coincide with text citations.

Parallel structure (parallelism)—a form of consistency—means related parts of sentences are written similarly. It expedites reading by using smooth connections to make sentences orderly and rhythmic. This requires similarity of tense (past, present, or future) and voice (active or passive) in paragraphs. Parallelism requires that words, clauses, and phrases have similar function and form within sentences. It demands consistency in headings, outlines, and series by using similar kinds of words and numbers in lists and bullets (Hopper et al. 1990). The sentence *The secrets to long, healthy lives are: don't smoke, moderate drinking, controlling your diet, get enough sleep and exercise, and staying home after dark* can be revised to put statements in similar form: *The secrets of long, healthy lives are: abstain from smoking, drink moderately, get enough sleep and exercise, and stay home at night.* Parallel structure is discussed in chapters 1 and 12.

Revision for Precision

Precision is sharpness, scrupulous detail, and definiteness. Precision differs from accuracy. Accurate statements can be expressed in imprecise terms and false statements can be related in precise terms.

Precision is essential in the materials and methods sections of research papers. These require exact concentrations, amounts, and sources of reagents so experiments can be duplicated. Once procedures have been precisely described, they can be discussed later in general terms.

Discretion is needed in use of precision. It is essential in reports and correspondence where names of people, places, dates, and organizations must be spelled correctly. The degree of precision in scientific descriptions varies with the situation. Reports to officials and press releases must be accurate but can be less precise than in publications (see chapter 1). Precision depends on personal writing styles. Accuracy, however, is always required in scientific writing.

Revision for Accuracy

Revision for accuracy is a crucial component of document improvement. It requires correct information, precise spelling and math, careful word selection, and correct referencing of direct quotes or controversial information. A reputation for accuracy underlies integrity and professional credibility. Fear of being wrong is one cause of writing blocks. Accuracy is initially addressed during revision and requires continual attention.

Revision for Spelling, Punctuation, and Grammar

Experience in revising, editing, and proofreading creates comfort with spelling, punctuation, and grammar. Without memorizing rules, writers can become astute at recognizing things that need correction.

Spelling

Chapter 10 offers advice about spelling. Most spelling questions can be answered by dictionaries or by replacing questionable words. Don't bother learning spelling rules. They are confusing. Make efforts to assure consistency in spelling of unique words and names. People's names are important to them (Carnegie 1998). If they remember nothing else they will remember if their name is misspelled.

Computer spell-checkers are valuable to detect misspellings, typos, and double word errors, but they may miss correctly spelled words that are incorrectly used.

Punctuation

Punctuation regulates reader speed by indicating pauses of various lengths with commas and semicolons. Colons indicate something important will follow and periods prepare readers for clarifications or new thoughts. Improving punctuation is part of revision. Its secrets are detailed in chapter 11.

Grammar

Grammar—more comfortably called the rules and recipes of writing correctness—is constantly changing. Despite former admonitions against splitting infinitives by placing modifiers between *to* and a verb, this is now acceptable. Many people find grammar confusing, and grammar books present conflicting interpretations of grammatical rules and definitions. When revising, search for anything that doesn't look or sound right. Be sure that:

- subject-verb-object order is used as much as possible;
- modifiers are placed near the words they modify;
- verbs match subjects in number, tense, and mood;
- pronouns match antecedents in number and gender; and
- parallelism is present in sentences, paragraphs, and lists.

If some of the above words are unfamiliar, consult the glossary, record their definitions on a card, and carry it with you until they are memorized. Chapter 12 summarizes grammar in its opening pages; the rest of the chapter serves as a reference.

Literature Citations

During drafting, some writers place asterisks where references will later be added. Others postpone inclusion of literature citations until drafts are completed. Inserting citations in text and developing reference lists can begin during revision.

Literature citations can appear as footnotes or references. Different citation formats are used in various disciplines. Determine the reference requirements of journals or publishers before proceeding too far with a project (Alred et al. 2003).

During revision, insert essential references and assure exact correspondence of entries, dates, and spellings between text citations and bibliographies. This can be accomplished by spelling out authors' names and dates in the text while a partner checks them in the reference list.

New, controversial, or recently updated information is usually referenced. It is also critical to assure that quoted or paraphrased statements are credited to their authors. If lengthy quotes, pictures, charts, or graphs come from copyrighted material, the author's or publisher's permission should be obtained. Excessive references are distracting. Many publishers suggest minimal references to save space and expedite proofreading. Two methods of literature citation are numerical and author-date systems.

Numerical citations list references by inserting parenthesized or superscripted numbers in the text in their order of appearance. They save space and are less distracting to readers than the author-date system. Numbering systems on computers simplify adding new references.

The **author-date system**—achieved by placing the author and year of publication in parentheses in the text—is favored by many journals and book publishers. It permits readers to identify authors and dates of cited material without consulting the reference list. Citations are listed alphabetically in the bibliography by the senior author's last name.

Electronic Citations

Referencing electronic material is challenging because the Internet is constantly changing. Material is often deleted; thus the retrieval date is essential. Some websites promote personal agendas, so it is preferable to quote refereed journals. When quoting information from the Internet, record the author, the name of the document, the contributing organization, the website address, and the date retrieved.

Final Revisions

After several stepwise revisions and layaway periods have been completed, a final revision is needed. Be sure that specialized words are initially explained, acronyms are initially spelled out, and changes in subtitles appear in the table of contents. The number of needed revisions is usually underestimated. Really good documents are revised repeatedly and improve each time.

Remember, the revision order presented here is only one approach. Revision sequences depend on each writer's personal comfort. After multiple revisions and layaways, a document is ready for editing.

CHAPTER 6

EDITING

Introduction

Editing smoothes, polishes, and refines documents. It follows multiple revisions and involves little moving of sentences or paragraphs. Authors usually edit their own correspondence and manuscripts before submission. Writers must be familiar with editorial strategies.

The Strategies of Professional Editors

Professional editors improve papers and books by upgrading readability and overseeing a publisher's style. Before starting, they usually read documents to get an overview of the message and its conformance to publishers' standards. They rarely praise documents. Most do a great job and, like authors, their skills improve with experience. Classic references for editors are *The Chicago Manual of Style,* 15th Edition (University of Chicago Press 2003), *Scientific Style and Format* (Council of Scientific Editors 2006), and *The Handbook of Technical Writing* (Alred et al 2003).

Editors return manuscripts with comments between the lines of double-spaced manuscripts or with notes on stick-ums. If they edit electronically, suggestions are identified by insertions in a colored font. Computer editing programs produce printouts showing the original document in reduced size, leaving an enlarged right margin for comments. Professional editors search for consistency of spelling, capitalization, punctuation, and grammar. They check to see that headings are consistent and that:

- paragraphs have opening sentences, nonredundant themes, and closing sentences;
- abbreviations and acronyms are written out when first used;
- literature citations are correctly spelled and dated in reference lists;
- unusual or technical terms are initially defined; and
- dangling phrases, clauses, or sentences are corrected.

After reading editors' comments, it is best to set manuscripts aside before responding or reworking them. When authors are aware of the issues targeted by editors, they can address them before submission.

Editing Your Own Work

Editing occurs every time we change a manuscript. Most people have some good editing habits. The journey to writing excellence will polish these skills. Editing resembles revision but involves less rearrangement and more focus on words and sentences. Presumed perfection is a good starting line for the editing marathon. Editing our own material is tricky because we tend to see what we think we wrote rather than what is actually there. That tendency is partially overcome by a layaway period after the final revision.

Editing can be done on computers, but some authors and professional editors choose to edit on hard copy. Use whichever you prefer. Your preferences may shift with time. Some people miss too much when editing long paragraphs on computer screens.

Experienced writers draft and revise manuscripts ahead of deadline and save the editing until traveling. Time passes rapidly when you're editing. Be sure to bring the project folder, a paperback dictionary, and your revision checklist. If your writing notebook still lacks a stepwise revision list, prepare one now. As a starter, use this list from chapter 5 and edit your manuscript carefully for:

- content, organization, and flow;
- word choice, placement, and spacing;
- the active instead of the passive voice;
- brevity, clarity, consistency, precision, and accuracy; and
- punctuation, spelling, and grammar.

This sequence can serve as an editing guideline until you adjust it to your personal preferences. You may wish to add items from professional editors' checklists. Eventually you will develop your own checklist. Some of your items will probably include points put forward by DeVries (2002), Elliott (1997), and Hodges et al. (1998).

An Author's Editorial Checklist

While most people are not skilled editors, they can do an excellent job at polishing correspondence and manuscripts. Editing checklists should include each item discussed below.

Edit to Upgrade Content

Editing for content determines that crucial points, illustrations, charts, data, and references are included and that non-pertinent information is deleted.

Edit for Format Compatibility

Format refers to the size, shape, layout, order, and appearance of documents. Many journals require standard formats. Some businesses and government agencies have standard formats for trip reports, budget requests, annual reports, and minutes of meetings.

Research journals contain instructions for authors that describe a format comprising an abstract, an introduction, a literature review, and sections on materials and methods, results, and discussion. These must be followed. Some journal or book publishers request double-spaced submissions which they mold to their specifications. Formatting also includes selection of type size and style, and the use of italics, underlining, and bold type. These all must be consistent throughout documents and compatible with publisher styles.

Edit to Upgrade Organization and Flow

Editing for organization and flow addresses questions such as:

- Is there an obvious introduction in the opening paragraph?
- Do paragraphs detail information suggested by the title and introduction?
- Does each paragraph have introductory, amplifying, and concluding sentences?
- Do sentences contain subjects, verbs, and objects in that order? and
- Does the abstract or conclusion summarize the text?

If these questions are answered affirmatively, move on. If there are still problems, address them before considering style.

Edit to Upgrade Style

Journal and book publishers require authors to follow their styles. Within those parameters, writers can invoke personal preferences as long as they use clear expression and proper grammar.

Edit to Upgrade Clarity

Clarity is improved by replacing weak words with specific terms and by shifting from negative to positive approaches. Search for passive-voice sentences that could be changed to the active voice. Look for words or phrases in lists that can be made clearer with parallel construction, and look for sentences needing more logical order.

Edit For Brevity

Eliminate wordiness, long strings of words, pretentious or non-specific words, and redundancies. Go on an adjective hunt to remove unneeded modifiers. If there are sentences and paragraphs that repeat the same word, replace redundant terms with synonyms.

Edit to Upgrade Consistency

As you edit documents for consistency, see that the type, style, format, spelling, lists, headings, subheadings, and references are similar throughout.

Edit to Upgrade Accuracy

Recheck the spelling of names and places, the accuracy of arithmetic, and the correctness of dates. Then make sure that all citations are correctly entered in bibliographies. For more on accuracy, see chapters 1, 5, and 7.

Edit to Upgrade Precision

Recall the distinction between accuracy and precision, and increase precision by replacing words like *large*, *numerous*, and *multiple* with exact values. Also replace words like *reagents*, *feed additives*, and *antibiotics* with names of specific compounds.

Edit to Upgrade Spelling

Use the computer spell-checker and pay particular attention to names, scientific terms, and commonly misspelled or misused words (see chapter 10).

Edit to Upgrade Punctuation

Check for appropriate use of periods, commas, colons, and semicolons (see chapter 11).

Edit to Upgrade Grammar

Without claiming grammatical expertise or memorizing definitions, edit documents for conformance with the basic rules of writing by checking for: agreement in person and number between subjects and verbs; agreement in case, person, number, and gender between pronouns and their antecedents; adherence to subject-verb-object order; and proper tense, number, person, and voice of verbs. Also check for:

- run-on sentences;
- incomplete sentences punctuated as if they are complete;
- correct placement of modifiers;
- appropriate capitalization; and
- consistency and parallelism.

If any of the above rules sound unfamiliar, look them up in the glossary or read about them in chapter 12.

Edit for Consistency of Headings, Fonts, and Spacing

Selection of font (type size and style) is not complicated. It is best to choose commonly used font sizes and styles and be sure they are consistent throughout a document. They can change in multi-authored manuscripts or when text is copied from other documents.

Edit Abbreviations, Acronyms, and Initialisms

Abbreviations, acronyms, and initialisms are shortened forms of words. They make sentences more compact. Their increased use responds to people's wishes to use monosyllables whenever possible (Maas et al. 2000). When used correctly and consistently, they make writing more readable. **Abbreviations** usually use the first few letters of words but may comprise any letters in a word. Their misspelling can confuse readers. Unless abbreviations will be readily recognized by all readers, they should be avoided.

Acronyms and initialisms arise from the first letters of multiple-word names. **Acronyms** create pronounceable words like AIDS (*a*cquired *immuno*deficiency *s*yndrome). **Initialisms** are collections of letters that are not readily pronounced as words but are verbalized as a series of letters like FDA (Food and Drug Administration). Acronyms and initialisms serve the same purpose and the distinctions (ease of pronunciation) are academic. Their letters are not separated by commas or periods. Acronyms and initialisms are made plural with a lowercase letter *s* and are usually made possessive with an *apostrophe s*.

Except for commonly used words like radar (radio detecting and ranging) or laser (light amplification by stimulated emission of radiation) whose meanings have been lost to time, the components of acronyms and initialisms must be fully spelled the first time they appear. Some scientists say people who don't know the acronyms of a discipline shouldn't be reading its materials. That contention overlooks translation difficulties; creation of abstracts for newsletters; and the interests of officials, legislators, or the media.

When editing, examine shortened forms carefully for initial explanations and proper usage.

Editorial Marks

Whether or not you edit the work of others, you need familiarity with editorial marks. They resemble the proofreaders' marks discussed in chapter 7. They eliminate the need to write out instructions and are convenient when editing hard copy. To use these standard marks, you:

- circle words or sentences that need movement and draw a line to their new location;
- place a caret (^) at the new location of moved or inserted material;
- change word order with a roller coaster mark (~);
- cross out deleted words with the hook-like delete sign;
- use the close-up sign to close spaces; and
- write stet—meaning save existing text—beside erroneous corrections.

If you don't already use these marks, start today. Editorial marks are found in dictionaries. They will soon become a habit.

The techniques of editing resemble those used in revision. Editing, however, requires more focus because the document is approaching perfection.

Editing Someone Else's Work

Traditionally, editing involves evaluation and improvement of other people's writing. There are so few excellent writers in some fields that your reputation will generate invitations to edit papers and serve on editorial boards.

If colleagues ask you to edit a paper, consider the pros and cons. Tell them that most writers avoid asking friends to edit their work because critical commentary can challenge friendships and freelance editors are available.

You can grow as a writer and can learn by critiquing the work of others. If you try it, make only suggestions that significantly improve documents. People take their writing seriously and may interpret suggestions as personal attacks. Therefore be diplomatic, lavish in praise of good ideas, and cautious with criticisms. Be sure you are right. Phrase comments as questions rather than criticisms. Editors should be constructive and avoid nit-picking about style and word choice. Explain that they are free to use or discard your suggestions. Then:

- ask them to lay away the document for a few days and reedit it before giving it to you;
- request a double-spaced hard copy;
- immediately duplicate it so you have several clean copies;
- clearly print questions, comments, and suggestions; and
- consider green instead of red ink to indicate support rather than criticism.

After these preliminaries, edit their work as if it were your own. Make sure all questions and comments are legible.

If you choose to edit electronically, print out the marked copy and return it with unmarked copies of the original. After finishing, prepare a positive cover letter that begins with praise and indicates that the suggestions are your personal opinions, and they are free to use them or reject them. Tell them you treated it as if it were your own and you are rough on yourself.

If you present your comments diplomatically and positively, you can develop wonderful professional relationships. If you are overly critical or impose your personal whims, you can lose friends.

Using Expert Advice

Some writers use dictionaries and writing books to help with grammar, punctuation, and spelling. Colleagues can advise you about scientific logic and accuracy. Imagine them looking over your shoulder as you edit, but never let anyone see first drafts. Rework drafts repeatedly before seeking advice or you'll be shattered by the comments. Helpers must be chosen with care and you should be prepared for insulting comments. Consider hiring freelance editors rather than involving friends.

It is valuable to have other people edit important documents. Second parties can identify ambiguities and lapses that writers overlook. Even though every document contains items worthy of praise and the tiniest commendation can inspire writers, accolades are unlikely to be forthcoming from editors. Their job is to raise questions and identify areas needing improvement.

Editing Summarized

Editing is the home stretch on the road to perfecting documents. It requires far more time than most writers devote to it. Time and energy devoted to editing will yield substantial dividends. With practice on your own work and study of key points, editing will become fun. Editing success will stimulate you to delve more deeply into proofreading, spelling, punctuation, and grammar. These are discussed in the chapters that follow.

CHAPTER 7

PROOFREADING

Introduction

Proofreading—the final correction before printing—focuses on perfection. It is a non-writing search for errors that survived revision and editing. Proofreading corrects typographical errors, misspellings, omitted capitalizations, misplaced punctuation marks, and grammatical flaws. It is best done with line-by-line reading of hard copy with unread text covered. Proofreading pays dividends in writing excellence.

A Three-Stage Checklist for Proofreading

The Handbook of Technical Writing (Alred et al. 2003) provides a three-stage proofreading checklist. Stage one seeks to improve consistency of headings, formats, fonts, and spacing. Stage two checks for accuracy of grammar, punctuation, abbreviations, capitalizations, spelling, literature citations, tables, and charts. Stage three is a final check of the author's goals, readers' needs, layout, and design.

Proofreading for Typographical Errors

Some proofreaders conduct a separate search for typos that have survived revision and editing. Some typos occur during the cut-and-paste process. Most result from missing a key, hitting the wrong key, or hitting a key twice. Others are spacing, double word errors, and omitted words.

Proofreading for Capitalization

Appropriate capitalization of names, cities, and countries—or its absence—is readily recognized. Dictionaries indicate words needing initial capital letters. Capitalization is detailed in chapter 11.

Proofreading for Accuracy of Tables and Charts

Figures, tables, charts, and graphs can contain errors that may be pointed out—to the embarrassment of authors—in letters to the editor. Books and meeting proceedings lack opportunities to acknowledge errors in subsequent issues. All numbers must be carefully checked against the original data.

Proofreading for Cut-and-Paste Errors

Computerized copying, cutting, and pasting simplify manuscript preparation but can cause difficulties. Redundancies result when moved material is copied rather than cut. This is detected when someone recognizes familiar wording and finds its original location. Other cut-and-paste errors occur when modifiers are left out of moved material or when the tense of the original passage is inappropriate in its new location.

Proofreading for Punctuation

Proofreading for punctuation requires a diligent search. Reading aloud is a good way to check punctuation because periods, commas, colons, semicolons, dashes, and parentheses indicate reader pauses or insertions. Chapter 11 reviews punctuation.

Proofreading for Spelling

Spelling shapes readers' judgments of author competence. Many writers and most editors conduct a separate read-through for misspellings that survive spell-checkers, for misuse of correctly spelled words, and for assurance that acronyms and initialisms are initially defined. Reading documents backwards is one way of proofreading for spelling. Chapter 10 has more hints on spelling.

Proofreading for Grammar

Grammar should be in good order after revision and editing. If a sentence still sounds odd, it may have escaped previous examinations. This is common when only the author reviews documents.

Proofreading is complicated by grammatical rules and definitions. These are best learned by experience. Attempting to master grammar in the abstract can be frustrating. If the computer identifies a problem, seek an explanation and make appropriate corrections. Another approach is to have a colleague read something that seems awkward and tell you what it is intended to say. Then rewrite it using subject-verb-object order. Chapter 12 reviews grammar.

Proofreaders' Marks

Proofreaders cross out, underline, or circle text needing changes. The needed adjustments are identified in margins with marks that indicate deletions, insertions, capitalizations, or other changes. Get a list of proofreaders' marks from a dictionary and practice using them.

Developing Effective Proofreading Skills

Proofreading expertise is simple to develop and leads to writing excellence. Like athletics and card playing, writing skills grow with experience. Errors found during proofreading are learning opportunities. Some proofreading suggestions are:

- lay away documents for two days before proofreading;
- proofread in a place free of distraction;
- go slowly and carefully—word for word—over hard copy;
- cover everything except the line being examined, and evaluate the position, meaning, and spelling of every word;
- proofread titles, headings, and subheadings carefully; and
- practice proofreading and applying proofreaders' marks on a variety of documents.

Enhancing proofreading skills is an effective step on the journey to writing excellence.

CHAPTER 8

DOCUMENTS NOT INTENDED FOR PUBLICATION

Introduction

The quality of not-for-publication documents signals professional competence. Email, memos, letters, press releases, class notes, committee reports, speech abstracts, and grant proposals all require different formats and provide opportunities to upgrade writing skills. *Webster's Business Writing Basics* (Merriam-Webster Incorporated 2001) offers examples of various kinds of correspondence. Prior to sending, examine these documents for appropriateness for all possible recipients and determine who should be copied.

Electronic Mail Messages

Electronic mail (email) reduces telephone calls and permits address storage, rapid messaging, and informal conversations among friends and colleagues. Its simplicity can lull users into carelessness and sabotage careers.

Many email addresses are unrecognizable. When recipients don't recognize the senders, messages are often deleted. This is avoided by putting the sender's name in the subject box with the title or by making the title so specific that recipients know exactly who sent it. Replies or forwards can also be deleted unless senders' names are obvious.

The ease of forwarding email means you never know who will receive them, so be careful what you write. Emails are best if short, confined to a single topic, and sent after layaway and careful proofreading. Write all emails as if they were going to a prospective employer. Here are suggestions for effective email:

- clearly identify yourself and your topic in the subject box;
- avoid using email for initial contacts with unknown individuals;
- if there's any hint of formality in an email message, use a letter instead; and

- if the message contains new material, indicate it is your opinion and not organizational policy.

Judiciously used email can be a windfall. Improperly used, it can cause a hard fall.

Memorandums

Memorandums (memos) are used for nonsensitive interoffice communications. They are less formal than letters. Their quality signals the sender's competence. The usual format for memos is:
- To:
- From: Sender's name, typed and initialed
- Re: or Subject:
- Date:
- Text:

Memos should be short, clear, to the point, and used when you know the recipients personally. Effective memos require descriptive titles, clear introductions, short statements about a single issue, and closing summaries. Hastily written memos can tarnish the images of capable individuals.

Facsimiles

A facsimile (fax) transmits previously printed matter like order blanks and account sheets. They are accompanied by fax transmission forms containing the number of pages, and the name, address, phone, fax, and email addresses of senders and recipients. Offices need someone to check incoming faxes, staple them to cover sheets, and deliver them to addressees' desks. If important material is faxed, the recipient must be notified so it is picked up promptly. Unexpected fax transmissions pile up and get lost.

Letters

Letters are more formal than memos and email. They set the stage for possible future relationships. Letters should project the writer's professionalism, reflect organizational pride, be sent on quality paper, and be brief. If they require multiple pages, there should be at least three lines of text on the last page. Letters project an image. They require careful revision, editing, and proofreading. Here is a recommended format for letters:

- letterhead including complete address;
- date right-aligned two or more lines below the letterhead;
- salutation with name (or Dear Sir or Madam) and address of recipient separated by two blank spaces above and below and followed by a colon;
- text including introductory, clarifying, and closing sentences;
- complimentary closing followed by a comma;
- signature block with name and full title of sender; and
- enclosure notices and names of persons copied.

Some format flexibility is permissible. Various styles for letters are presented by Sabin (2001).

Expert Committee Reports

Government agencies and legislative bodies often assemble expert committees to analyze issues, develop consensus-based reports, and recommend actions. These reports are often read by laymen and can be distorted unless they are clearly written in nonscientific terms. Ideally, the initial paragraph is a summary.

Expert committee reports must be carefully revised and heavily edited to reduce bias; achieve consistency (often missing in multiauthored papers); and be suitable for release to the media, who may use the summaries as press releases.

Press Releases

Press releases are candid statements about organizational programs and policies. They help cultivate media and gain public support. They should be brief, concise, and free of exaggeration, deceit, advertising, or promotional material.

Press releases are underutilized due to the challenges of achieving intra-organizational unanimity, and concern about media editing. There is no assurance that they will be published verbatim. A format for press releases is:
- organization's letterhead;
- a heading, P R E S S R E L E A S E, double-spaced in caps at top of the page;
- a title line with the TITLE (capitalized);

- a dateline with the CITY (capitalized) followed by the date; and
- a double-spaced body comprising an introductory sentence, one to three qualifying sentences, and a concluding statement.

Press releases should not exceed one typewritten page. They are excellent devices for developing relationships with the media. Newspaper reporters and magazine editors are often desperate for material and will use clearly written releases.

Routine, nonurgent, and carefully crafted press releases that are circulated within organizations will update everyone about programs and priorities, prepare them for external queries, and upgrade organizational communication.

Class Notes

Many teachers prepare handouts so students can focus on class discussions. Some institutions require class notes. This expectation presents challenges in courses with multiple lecturers.

Class notes may be skeletal outlines, lists of topics with blank space for notes, brief frameworks with acronyms and complicated words defined, printed PowerPoint presentations, or summaries of material discussed in class. When class notes are updated annually they gradually become textbook outlines.

Some colleges require class notes. Department chairs and deans may want copies. This request can cause panic if received unexpectedly, just as classes start.

Trip Reports

Some organizations require written reports with travel reimbursement forms. Word gets around if these requirements are not enforced. An unexpected demand for trip reports can trigger writing blocks. Trip reports present opportunities to practice writing. They contribute to organizational communication and reinforce the writing skills of their authors.

Start summarizing each trip and you'll soon have a trip-report template stored on your laptop. It will include a succinct summary, the date and purpose of the trip, names of attendees, things discussed and accomplished, and suggested actions.

Meeting Summaries

The comments about trip reports also apply to meeting summaries. Colleagues are eager to know about meetings and you may be one of the few individuals willing to share this information. People will appreciate and forward your reports. Meeting summaries that consider human sensitivities and confidentialities will reflect your improving writing skills and advance your career.

Issue Analyses

Written issue analyses are rare in organizations. Many issues are discussed but not summarized. Clearly written issue analyses record details that might otherwise be lost. They contain a description of the issue, its impact on the organization and society, and possible questions and actions.

Curriculum Vitae

Curriculum vitae are detailed descriptions of peoples' education, professional activities, organizational memberships, speaking engagements, and publications. CVs should be updated annually and whenever manuscripts are published or professional changes occur. CVs are reviewed by potential employers and promotion committees. They must be meticulously written, accurate, and list only degrees that have already been awarded. Publication lists are important components of CVs.

Biographic Sketches

A biographic sketch is a one-page abstract of a CV. Bio-sketches summarize professional activities. They are altered for speaker introductions, job inquiries, or queries to publishers. They reflect their author's competence and must be clear, succinct, and meticulously written.

Grant Proposals

Organizations that fund research require written proposals and usually receive requests exceeding available monies. Grant proposals determine projects selected. They must address the concerns of the granting agencies and adhere to format and submission requirements.

Successful grantsmanship depends on what you know; who you know; and how well you write proposals, thank-you letters, progress reports, and manuscripts. Grant acquisition requires that applicants know the current topics of public concern in their field, the modus operandi of granting agencies, the nature of available funds, and the people to contact for application information.

Successful grantsmanship requires timely project management, publications that acknowledge funding sources, and alignment with multidisciplinary or intra-institutional projects.

Writing skills largely determine grant-getting success. If the granting agency has a standard format for proposals, submissions must follow its guidelines, size limitations, and deadlines. Within that format, be sure to describe the questions addressed and how they will be answered, the instrumentation available and the apparatus needed, and the qualifications of the investigators. Clearly outline projects and convince review committees that the studies will positively reflect on their program.

Grant proposals usually contain an introduction, body, and conclusion. *The Handbook of Technical Writing* (Alred et al. 2003) details proposal contents as front matter, body, and back matter. The front matter contains a copy of the grant announcement, a cover letter, a title page, and a table of contents.

The body of grant proposals includes an executive summary, a statement of the problem, the rationale for the study, a review of existing literature, details of proposed experiments or surveys, itemized cost estimates, qualifications of personnel involved, facilities and equipment needed, and a conclusion. The body is followed by end matter containing biographic sketches of participating workers, appendices, a reference list, and a glossary.

The quality of writing is a factor in the success of grant proposals. They require multiple revisions, layaway periods, careful editing, and alert proofreading.

Becoming the Writer in Your Group

Your organization may have a correspondence-tracking system requiring documents to be sent through a supervisor. If so, send trip reports, issue analyses, and other papers electronically to your boss and copy your subordinates. Add a boldface caveat saying "Draft for Discussion: Not Organizational Policy" on the first page. This usually exempts messages from correspondence-tracking details, freedom of information requirements, and use in legal actions. If your documents convey useful information, are well written, positive, and free of character defamation and criticism, they will be appreciated and forwarded. If you are among the few report writers in your group, your reward will be challenging assignments, speaking invitations, and promotions. If you are a "loose cannon" who criticizes your organization and its management, please don't do it in writing.

Conclusions about Documents Not Intended for Publication

Care is required in submission of memos, letters, reports, and proposals. These documents establish a reputation for competence, organizational loyalty, and professional credibility. They provide the practice field for preparation of manuscripts for publication. Before they leave your desk, evaluate them as if they were going to a committee approving your promotion.

CHAPTER 9

DOCUMENTS FOR PUBLICATION

Introduction

Preparing and submitting documents for publication frequently gets inadequate attention due to deadline pressures. Successful authors meet the requirements of publications and submit timely and concise documents. Submission requirements of research and professional journals are described in their Instructions for Authors. If you follow their guidelines, they will usually refer your papers to editorial boards for consideration.

Some book publishers solicit titles from celebrities but won't look at other works unless they have requested them after viewing queries or proposals. Others accept books only from agents hired by writers.

Timely submission is essential. Experienced writers appreciate the time required to complete documents. They begin early. Short deadlines—sometimes due to delayed starts—produce shoddy documents. Every minute counts because time is needed for drafting, revising, editing, proofreading, and layaway periods. Therefore start early.

Publishable documents such as speech abstracts; graduate theses; journal, newspaper, and magazine articles; and books all have unique submission requirements.

Speech Abstracts

Meeting organizers often seek speech abstracts for distribution to attendees. These requests trigger speaker activity. Due to lead-time requirements, abstracts can differ markedly from the actual talks. When abstracts are submitted after meetings, they more accurately reflect presentations. Some associations won't pay travel expenses or honoraria unless abstracts are received. Even so, some speakers don't submit them.

Books of the meeting proceedings contain speech abstracts and are provided to meeting attendees. Generally, meeting proceedings are less rigidly edited than journal articles. While less prestigious than refereed publications, speech abstracts are an indication of competence and are listed as nonrefereed papers in CVs.

Graduate Theses

Theses required for graduate degrees are stressful and challenging. They must meet the requirements of both universities and graduate committees. Theses usually contain an introduction, a literature review, a description of materials and methods, a results section, a discussion, and a conclusion. They require extensive revision, editing, and proofreading. Begin drafting theses as soon as the topic is selected.

Scientific and Professional Journals

Publication in journals is essential for professional recognition. Research papers must report original studies and must be submitted to a single journal for exclusive use.

Academic tenure and promotion guidelines emphasize differences between refereed and nonrefereed journals. This distinction is fuzzy, but prestigious journals are recognized by their quality, their rejection levels, and the reputations of their editorial boards. Before submitting to professional journals, carefully review their requirements. If you have questions, call them. Journal articles must be repeatedly revised, edited, and proofread before submission.

Except for nonexperimental reports, journal articles usually have introductions, materials and methods, results, discussions, and conclusions. The introduction describes the current status of the topic. It provides background for the project, explains why the study is needed, and briefly mentions the approaches used. The material and methods section details procedures and describes all tests and statistical analyses. It contains references about techniques and footnotes detailing the reagents. The results section accurately describes the outcome of projects in the order outlined in the materials and methods. It reports findings, complications, or revelations. The discussion section explains the significance of findings, compares them with previous reports, and suggests additional studies. The paper closes with a summary and conclusions.

Nonrefereed journals are hard to find in most fields. Even less rigorous professional publications have editorial boards, but there are varying levels of strictness. Experts in the field usually know which is which.

Lengthy author lists should be avoided in journal articles unless every contributor worked actively on the project and its document-tation. Those who just assisted can be thanked in the acknowledg-ments. Research papers are detailed by Zeiger (1991).

Publishing for Reimbursement (Freelancing)

Requirements for reimbursement vary among publishers of news-papers, magazines, and books. Approaches for earning royalties are detailed by McCollister (1999). Publishing for pay often depends on publishers' responses to queries or proposals. If you are an employee of an organization and the message is on their behalf, publishers usually won't pay royalties. Some professional journals levy a page charge.

Freelance writing sounds appealing. Some hints for writing at home are presented by Shimberg (1999). Freelancing may provide a secondary source of income, but most who dream of writing for pay are disappointed. The few who succeed as freelancers offer combinations of unique subject matter, carefully written material, aggressive marketing, passion, patience, and dedication of time. Marketing is frequently the missing ingredient.

Marketing begins before you start writing. McCollister (1999) says if you are writing for dollars, you should:

- identify a need for books or magazine articles;
- study back issues of target periodicals to see what they have in your area;
- target specific audiences, find out what publishers want, and develop a product that addresses those needs;
- advertise yourself and your writings at meetings, conferences, and interviews;
- submit effective query letters and proposals;
- evaluate and adjust your marketing strategy periodically; and
- once an article is published, seek sequels, similar articles, or revisions in different publications and in other markets.

In his book *Writing for Dollars*, McCollister (1999) suggests reading everything with a "writer's eye" by considering how it could be rewritten from another perspective. His tips for freelance writers restate the need for concerted effort and dedicated writing time.

Ghostwriting

Ghostwriters are anonymous and receive no acclaim or criticism for quality or content of books, articles, or speeches. Semi-ghostwriters get acknowledgment in a byline or as a coauthor. Ghostwriters must continually network to make contacts with officials or celebrities. They need written agreements before handing over the work. Speakers and authors who employ ghostwriters expect them to remain anonymous. This may make it challenging to market these services.

Submissions to Newspapers and Magazines

Writing for newspapers or magazines can be emotionally rewarding and profitable. There is growing public concern about food safety, emerging and zoonotic diseases, the environment, and global tranquility. These and other issues offer opportunities for writers who are willing to explain them in understandable and unbiased terms. Those who work at it may find themselves writing syndicated columns and in demand as ghostwriters.

Preparation and Submission of Books for Publication

Mainline book publishers are declining and rapidly being replaced by print-on-demand (POD) operations (see below). When considering a book, make early inquiries about the interests of publishers. Acquisition editors can be met at publishers' booths at conventions. They will state their expectations. If you don't have a direct contact with a publisher's representative, send a query letter to determine their interest in your topic. Query letters are best addressed to a specific acquisition editor whose name—and its correct spelling—can be learned by calling the publisher's office. Some publishers won't deal with unpublished authors and will accept proposals only from literary agents.

When preparing books, initially insert the chapter and page number in the footer of each page to keep them organized during repetitive revision, editing, and proofreading. These can be deleted before submission if the publisher will be numbering the pages.

Book Queries

Query letters test publishers' interest and help them evaluate prospective authors' writing skills. Except for celebrities, most mainline publishers won't accept books without an invitation based on a query or proposal. Query letters are best kept to one page. If you are thinking of a book, prepare queries early. They are best written on your letterhead and accompanied by a self-addressed, stamped envelope (SASE).

Query letters contain the title, a brief description of the book, its potential audiences, and the estimated number of words (not pages). Word counts are available in Microsoft Word under Tools/Word Count. Queries should include a list of marketing channels, a review of competing titles, and a statement of how your book differs from existing works. Query letters conclude by asking if the publisher would like a proposal.

Book Proposals

Publishers' requests for proposals are merely a sign of interest. Proposals can occupy from three to seven pages. They illustrate your best writing style and describe: the book, why it is different, why you are qualified to write it, the number of words, a list of chapters, a sample chapter, the approximate completion date, and a list of potential buyers. After they review a proposal, publishers may ask further questions or send a contract or rejection letter. Many never respond. Major publishers reject many books that end up as best sellers.

Print-on-Demand Book Publishing

Print-on-demand (POD) publishers charge for producing books. They were once called vanity presses, implying—often incor-rectly—that they produce ego-based writings of low quality. Most POD publishers leave marketing to the author, as do many mainline publishers. This is challenging unless the book has a huge potential readership or a well-known author.

Marketing requires advertising and contacts with buyers, Internet marketers, bookstores, and libraries. Meetings, speaking invitations, and book signings provide opportunities to generate readers. POD publishers sometimes handle sales and shipping but some leave that to authors. Bookstores pay half-price or less, demand refunds for unsold copies, and often place self-published books in obscure locations.

Use care in choosing POD publishers. Check them out with colleagues in writing groups, with Better Business Bureaus, and by studying their websites. Publishing-on-demand, sometimes called "author-originated" book publishing, is described at and by InfinityPublishing.com (2007).

Multiple or Single Book Authorship

The pros and cons of multiple authorship must be considered before agreeing to coauthor books. Multiple authors add dimensions to books but complicate writing, editing, and revision. A senior author may be saddled with editorial responsibilities and authority to replace contributors who miss deadlines or deviate from the publisher's format. The hazards of such projects are variable deadlines; differing writing styles and computer programs; and disagreements about content, emphasis, and placement of subject matter. Multiauthored books have prolonged development times due to late submissions and complex editing. Some end up as hodgepodges and others never reach the publisher.

Think twice before agreeing to serve as leader of a multi-authored book. Your rewards will be credit as senior author and lots of headaches. Be aware that colleagues may disappoint you with late or absent submissions or chapters that lack clarity or deviate from the topic.

Jointly Authored Books

Agreement to write a book with colleagues broadens the expertise and scope of the work and can cement friendships. Each contributor adds challenges due to variations in style and deadline adherence. Multiple authors can also create philosophic differences and redundancies. When participating in joint authorship, it is important to know the senior author, the other contributors, the conditions of the agreement, and the distribution of royalties.

Book Chapters

Preparation of chapters in multiauthored books can be challenging. Be familiar with the authors of other chapters so you can contact them to avoid duplication. Don't count on senior authors to do this. They will be busy chasing delinquent manuscripts.

Single-Author Books

A single-authored book can be a major task. It requires lots of work and gigantic marketing responsibilities. However, a sole author can exercise personal choice regarding topic selection, emphasis, organization, and style.

Single-author books should be preceded by a search for competitive volumes and the need for superior products. They require broad knowledge, intense effort, and extensive literature reviews. Books don't just pop up, they evolve over the years. Book writing leaves legacies. It focuses and exhilarates careers.

Books emerge gradually from collections of class notes, journal publications, and speech outlines. Every written document should be stored on clearly labeled discs or CDs and hard copies. This collection can ultimately be a book.

Start now and you'll be glad you did.

Having a book as a secret target can narrow activities to a workable range and provide incentive to develop expertise in a privately staked-out niche.

The thought of writing a book may seem wild. It will become logical as your growing writing skills result in increased speaking invitations and a reputation as an expert. Reflect on visiting the offices of retired colleagues whose sole legacy was boxes and file cabinets full of unique information destined for trashing. Think about a book, begin collecting, and don't give up.

Unless books are required texts for large courses, few writers receive royalties adequate to repay the energy devoted to them. An honor, however, is to be introduced by a chairperson who waves your book and says "Our next speaker wrote the book on the subject."

Book markets are in transition. Mainline publishers are merging and being challenged by small presses and self-publishing companies that produce low-priced paperbacks.

Marketing Books

The changing book industry leaves authors largely responsible for marketing. Book marketing is a complex operation. It begins with choosing an attention-getting title and establishment of a reputation as an expert with public speaking engagements, radio and television interviews, book signings, and magazine articles. Unless you have the personality and energy to generate such fame, your book income probably won't justify the energy invested. If you do it to contribute to your profession, or leave a legacy, you will be adequately rewarded.

In today's competitive book market you will be the major marketing agent, so begin planning before-and-after-publication marketing activities. A short, dynamic title is essential. Design a front cover, back cover page, and binding that catch buyer attention. Develop a thirty- to forty-word back cover attention-getting blurb that can also serve as a thirty-second elevator speech. For example: "If you dread and postpone writing, this amazing book is for you. It serves as a stepwise pathway to writing excellence and can be skimmed, studied, or used as a reference. It will invigorate your career."

Develop a website, accumulate addresses of potential reviewers or customers, and prepare a list of people, libraries, and others to whom you will give a complimentary copy.

Manuscript Rejections

Manuscript rejections are everyday occurrences. They are the usual responses from book publishers and from prestigious magazines and journals. Book publishers receive more proposals than they can possibly publish, and reject most submissions. Rejection notices are disappointing and can cause anger (McCollister 1999). Recognize that the manuscript was rejected and not you. Best-selling books are often rejected 10 to 30 times before one publisher gambles on them. Journal rejections may be more amiable than those from book publishers. Some journal rejection letters indicate reasons for rejections, such as:

- the subject matter is inappropriate for the journal;
- the manuscript deviates from the required format; and
- differences with existing theories were inadequately explained.

Journal editors often suggest improvements. Whatever the reasons for rejection—explained, implied, or omitted—be assured you are not alone. Read rejection letters carefully and set them aside. Later, in concert with coauthors, address each comment, conduct a revision, and resubmit to another publication. Avoid negative reactions or angry responses. Thank publishers or editors who provided comments or suggestions.

Publishing Your Organization's Message

If an organization has a message, newspapers and magazines can be their spokespersons. Journalists can be met at class reunions, meetings, seminars, and short courses. They may accept your material without paying and print it as a space filler. Be sure that the message is well written and covers issues of public interest. It should be introduced with attention-getting information and be informative and unbiased. It must be free of advertising, regulations, jargon, or funding requests. You can publish your organization's message by providing the media with press releases and the permission to use them as needed. However, this can result in edited versions that differ from your copy and you may feel sabotaged. Regard this as a signal that your message could have been clearer and more convincing—not as a betrayal. Avoid responding. Rework it and try again. The media people think you are their friend. Keep it that way.

Some organizations have information specialists who are trained in writing but have limited technical backgrounds. They create documents and speeches for leaders on short notice and often have trouble getting help. If they find you cooperative and willing to provide information, or even willing to edit or write for them, they'll bond with you. This lets you participate anonymously in developing organizational policy. You may have writing skills but fail to get approval because the organization's correspondence tracking system clarifies everything to death, your views differ from organizational policy, or your superiors think they can do better (but never get around to it).

Blending Components, Applications, and Audiences

Effective writing requires consideration of all possible audiences. Understandability can be improved by initially defining complex words, abbreviations, and acronyms; by limiting jargon and specialized terms; and by using the fewest and simplest words possible.

The requirements of publishers are met by revising, editing, and proofreading to achieve their format and by using appropriate punctuation, spelling, and grammar. These efforts will yield concise narratives that convey integrity and credibility.

Conclusions Concerning Submitting Manuscripts

Manuscript submission culminates months of work. Your product should be:

- the best you can produce;
- devoid of pretentious language and criticism;
- inclusive of both sides of controversies; and
- carefully edited and proofread.

When these requirements are met, flow, clarity, and accuracy will result. Then your submissions will have a good chance of surviving review.

CHAPTER 10

SPELLING

Introduction

Spelling—the order of letters in words—is essential to reader-friendly documents. Writers are rarely praised for good spelling but readers quickly recognize errors. Spelling shortfalls can result from accents or dialects or may be baggage from school days. Discomfort about spelling is one component of writing blocks.

Spelling guidelines change over time and vary between countries. U.S. spelling rules are loaded with exceptions. Spelling excellence is achieved by conscientious effort, practice, use of dictionaries, careful proofreading, and by prudent use of computer spell-checker programs.

The Advantages of Good Spelling

Correct spelling helps achieve consistency, precision, and accuracy. Unlike speaking, in which spelling is irrelevant, spelling errors stand out in writing. Correct spelling in documents of all sorts indicates an author's competence.

Spelling ability varies among individuals due to brain types and the level of emphasis by parents, teachers, and writers. Many excellent speakers have trouble spelling. It is comforting to know that almost everyone misspells some words (Cazort 2002).

The Causes of Spelling Errors

Misspellings can arise from carelessness, use of commonly misspelled words, variations in pronunciation, incorrect word choices, and typos. Computer spell-checker programs defend against these causes.

Carelessness

Many misspellings arise during writing, editing, proofreading, or spell-checking—particularly if done hastily or in interruptive environments. These can be corrected by working in privacy, concentrating, and proofreading carefully.

Commonly Misspelled Words

Commonly misspelled words appear frequently. Writing and grammar books by the University of Chicago Press (2003), Alred et al. (2003), and Thurman (2003) list words that defy the rules and must be memorized. Some that appear in the writings of professionals are:

- aberrant, absorption, analytical, anesthetic, basically, bronchial;
- dependent, desiccate, diphtheria, effervescent, erroneous;
- fourth, hemorrhage, hereditary, idiosyncrasy, inconvenience, occurrence, pasteurize; and
- psychiatry, questionnaire, synchronous, and veterinarian.

Have someone read these words to you and write them out. Memorize the ones you misspell.

Variations in Pronunciation

There is not always a direct relationship between pronunciation and spelling (Steinmann and Keller 1999). Nevertheless, regional variations in pronunciation can undermine spelling.

Incorrect Word Choice

Misspellings can result from incorrect choices between similar words. If spelled correctly, similar sounding words with different meanings can be overlooked by spell-checker programs and can significantly alter the message. Easily interchangeable word pairs include:

- *absorption* (noun) means the process of drinking in, soaking up, or being ingested; and *adsorption* (noun) is the process of collecting on surfaces, as with condensation.
- *compare* (verb) means to note points of similarity or difference; and *contrast* (verb) means to note differences.
- *imply* (verb) means suggesting or hinting by speakers; and *infer* (verb) means concluding or understanding by listeners or readers.

- *supplant* (verb) means to take the place of; and *supplement* (verb) means to add to.

There are many more. Write out, memorize, or sidestep combinations that give you trouble. As seen from the list, spelling and word selection are related.

It is important to know definitions as well as correct spellings. Dictionaries present different, and sometimes conflicting, definitions of the same word. More extensive lists of commonly confused words appear in the *Random House Webster English Language Desk Reference* (Maas et al. 2000) and in grammar and writing books.

Typographical Errors

Many misspellings result from typographical errors that can be recognized and corrected during computer spell-checks, editing, and proofreading (see chapter 7). Typos are easily overlooked.

Spell-Checker Programs

Computer spell-checker programs usually work in concert with grammar checkers. They identify misspellings and typographical errors but may not detect correctly spelled words that are incorrectly used.

Spell-checkers will not resolve all problems. You must personally check spelling by carefully proofreading documents. Watch out for spell-checker accidents. Errors can be permanently embedded in spell-checker programs if incorrect spellings are entered into dictionaries. Spell-checkers are also discussed in chapters 2, 6, and 7.

Spell-checker programs are excellent aids but spell-checked documents must be carefully reviewed for proper spelling and word selection.

Elaborate Spelling Rules

There are some simple spelling rules but they are outnumbered by those written in grammatical jargon and complicated by multiple exceptions. Even the "I before E except after C" rule has exemptions, including caffeine, codeine, foreign, seizure, and tracheitis; proper nouns naming species like Leishmania and Neisseria; and names of people like Alzheimer and Feingold.

Dictionaries are the best source of spelling information. If you choose to become a spelling scholar, detailed guidelines are available in books (Shertzer 2001, University of Chicago Press 2003, and Beckett-McWalter et al. 2002). They address making words plural, indicating possession, forming compound words, adding prefixes and suffixes, and using abbreviations and acronyms.

Unless you majored in English, it may be frustrating to try to memorize spelling rules. Instead, you may choose to memorize the words you frequently misspell, refer to dictionaries, revise and edit with care, and develop meticulous proofreading techniques.

Writing Numbers

Rules for writing numbers are less complicated than spelling rules. In sentences, only single-digit numbers are usually spelled out as words and numbers 10 and higher are expressed as numerals. Numbers that begin sentences, regardless of their size, are written out as words, with the first letter capitalized. When several numbers appear in a sentence, they are all treated the same regardless of size.

Prefixes and Suffixes

An important part of spelling is to assure that prefix- and suffix-containing words are correctly chosen to convey the author's intentions. Prefixes are letters placed before root words to adjust their meaning or number. They usually create new words. They are helpful when correctly used but can be confusing if used inappropriately. Some commonly used prefixes are:

- *ante* means before (antemortem);
- *contra* means against (contradiction);
- *ex* means out of or away from (exteriorize);
- *pro* means before (projection); and
- *post* means after (postpartum).

Prefixes are attached directly to root words unless the prefix and the word to which it is attached both begin with vowels or if the root word is a proper noun that is capitalized, such as anti-American.

Suffixes are letters placed at the end of root words to create new words. Unlike prefixes, suffixes indicate the context in which words are used in a sentence. There are noun, verb, adjective, and adverb suffixes and they indicate the part of speech in which the attached word is used. Words containing suffixes must be carefully examined to assure their appropriate use. Some commonly used noun suffixes are:

- *ectomy* means removal of (appendectomy);
- *ism* means the quality or state of (dwarfism);
- *itis* means inflammation of (gingivitis);
- *phobia* means fear of (hydrophobia); and
- *plasty* means pasting to (dermatoplasty).

A frequently overlooked aspect of editing and proofreading is the assurance that suffixes are appropriately used to express authors' intentions.

Managing Acronyms and Initialisms

Proper spelling of acronyms and initialisms requires careful proofreading. It is customary to add the lower case letter *s* without an apostrophe to make them plural (Alred et al. 2003). Careful proofreading (see chapter 7) is essential to achieve consistency in use of acronyms and initialisms.

Workday Spelling Rules

Recognizing the complexity of making spelling decisions, Douglas Cazort (2002) presented working rules for selecting proper spellings. They include:

- write first, edit later, and ignore spelling and other details until drafting is complete;
- look up correct spellings only during editing and proofreading;
- use spell-checker programs after drafts are finished; and
- have a good speller proofread your writing.

If consistently applied, these approaches can probably do as much for your spelling as memorizing 50 or 60 exception-laden rules.

National Differences in Spelling Practice

British and some Canadian spellings, particularly for words ending in *er* or *or*, may differ from American usage (Austin 1999). Some differences between American and British spellings include:

- anemia/anaemia;
- flavor/flavour
- hemorrhage/haemorrhage;
- odor/odour; and
- tumor/tumour.

Once such differences are recognized, writers, editors, and proof-readers are comfortable with them.

People from some English-speaking areas and most non-English speakers may have difficulty understanding Standard American Written English. If these people are among your potential readers, care is needed to minimize colloquialisms, clichés, idioms, and jargon, and to initially spell out acronyms and initialisms. Some experts feel that dramatic changes in spoken vocabularies will make American English incomprehensible to people from other English-speaking countries by 2025.

Achieving Consistently Good Spelling

Good spelling is easily achieved by authors who: write in words they use in speech, practice spelling problem words, consult dictionaries and thesauruses, proofread carefully, and apply spell-checker programs. Once we acknowledge its contributions to writing excellence, spelling can be addressed aggressively in a manner comfortable to each of us.

Write Mostly in Words Used in Speech

People can usually spell words they use in conversation. When writing, we often shift to impressive words that are more easily misspelled. Many spelling problems can be overcome by limiting our writing to words we commonly speak. Examples of commonly spoken words and their easily misspelled written substitutes (Andersen 2001) are:

- *trouble* written as *inconvenience*;
- *show* written as *reveal* or *demonstrate*;
- *later* written as *subsequently*; and
- *stop* written as *terminate*.

Replacing such words with their common counterparts improves clarity, brevity, and spelling.

Practice

As in many skills, practice and repetition are essential to improve spelling. Make a list of words you misspell, keep it in a pocket or purse, and study it until you master them all.

Use of Dictionaries and Thesauruses

One way to improve spelling is to consult dictionaries and thesauruses. They provide spelling guidance, clarify meanings, and list synonyms. There is constant change in word usage, so seek dictionaries that are less than five years old (Alred et al. 2003). As a rule, newer dictionaries are better. Definitions of words may differ from dictionary to dictionary. Choose the word that best conveys your intended meaning but don't use it more than twice in the same sentence or more than three times in a paragraph.

There are many types of dictionaries. Abridged paperback dictionaries are adequate for common words. Serious writers stock their writing hideaways with recent editions of unabridged dictionaries with a minimum of 300,000–500,000 entries. These provide detail, alternative word uses, synonyms, and acceptable spellings but are too large to fit in briefcases. Software programs and online dictionary sites are convenient, as well.

A thesaurus helps you to find alternative word selections and to check spelling and definitions. A good addition to your library is a two-in-one volume containing thesaurus and dictionary entries on the same pages (Agnes 2002).

Editing and Proofreading

Spelling is a bona fide point of focus during editing and proofreading because new words appear each time a document is reworked. Proofreading focuses on spelling, and many proofreaders conduct a separate reading devoted exclusively to spelling.

Other Spelling Strategies

Spelling strategies involve identifying and attacking the causes of misspelling that are most applicable to your own writing. One technique is to estimate your spelling prowess by taking a spelling test. This is easily accomplished by having a friend read a list of commonly misspelled words. Make a list of the ones you use and work on it until you've memorized them all. Such lists are widely available (Maas et al. [2000], Steinmann and Keller [1999], and Shertzer [2001]). Comprehensive lists are in *Random House Webster's Handy Grammar, Usage, and Punctuation* (Random House Publishing Company 2003), and *The Only Grammar Book You'll Ever Need* (Thurman 2003). Other tricks for spelling improvement include:

- activating the spell-checker program and correcting highlighted words without consulting the selection list to improve your spelling and help identify your most common typographic errors;
- highlighting words you have looked up in your dictionary; and
- maintaining a list of correct versions of your misspellings for study and reference.

These methods yield immediate feedback.

A Checklist for Improving Spelling

DeVries (2002) presents a checklist for minimizing spelling errors. It includes systematically screening manuscripts for:

- typographical errors;
- commonly misspelled words;
- consistent spelling of words with variable spellings;
- correct use of numbers as words or figures; and
- consistent use of abbreviations, acronyms, and initialisms.

These steps, applied after completing spell-checker programs, will improve spelling.

CHAPTER 11

CAPITALIZATION AND PUNCTUATION

Introduction

Capitalization and punctuation help readers pace themselves. Capitalization focuses attention on titles, important words, and new sentences.

Punctuation signals length of pauses and stops by using commas, semicolons, colons, and periods. It also creates plurals and possessives with apostrophes and signals insertions with parentheses, brackets, dashes, slashes, and quotation marks (Alward and Alward 2000). Less commonly used punctuation marks are exclamation points, question marks, ellipsis points, and braces.

Punctuation lets writers express things that pauses, body language, facial expressions, and volume changes achieve in speech.

Capital Letters

Capitalization changes the first letter of words to their large form (upper case). It is used when words need special attention, have special meaning, or begin a sentence. Capital letters begin names, titles, and ranks. Capital letters also initiate the spelling of specific places, agencies, bureaus, companies, institutions, streets, buildings, states, territories, countries, days of the week, and months of the year.

Periods

Periods, the familiar dots ending sentences, also are called end stops or full stops. They provide a written counterpart of the pauses that speakers use to introduce new thoughts. Periods also separate initials in names, serve as decimal points, end sentences, and create abbreviations. Periods were once used to separate letters in multiple word abbreviations but this use has been largely replaced. Thus, New York's abbreviation became NY rather than N.Y. Acronyms such as AIDS and initialisms like HIV likewise are devoid of periods. If sentences end in abbreviations that have a final period, a second period is not added.

When sentences ask questions or make demands, periods are replaced with question marks or exclamation points. Periods are usually placed inside quotation marks. Three consecutively spaced periods constitute an ellipsis (…) and indicate pauses or omissions.

Periods receive special attention during proofreading. It is essential to see that single periods end all sentences, serve as precisely placed decimal points, and follow the last letter in classic abbreviations. They are not preceded by a space. Careful use of periods easily becomes a habit that creates easy reading.

Commas

Commas indicate divisions or separations between words and phrases. They establish reading rhythm. Comma use is flexible and changes over time. Once, series of words were all separated by commas except for the last two that were separated by conjunctions like *and* or *but*. Now, commas also precede the closing conjunctions. Newspapers omit the last comma to save space. If editors recommend a comma, let them have their way unless it changes the meaning. Commas are used to separate:

- similarly constructed words or phrases in a series;
- introductory phrases from the main portion of sentences;
- elements of dates, addresses, and salutations; and
- people's names from their titles.

Other uses of commas are to: precede coordinating conjunctions (*and, but, for, or, so,* and *yet*) behind series of words and to set off long appositives—words that explain who or what—from the subject of a sentence, as in: Bill, *the department chair who just arrived,* said commas are confusing.

With some exceptions, most spoken pauses require a comma when written. Excess commas can distract readers. When commas appear to interrupt word flow, leave them out. Situations in which commas may be omitted (Shertzer 2001) are:

- when modifying terms immediately follow a word, as in: Pat *the radiologist* died.

- when two words are connected by a conjunction like *and* or *but*;
- when all words in a series are connected by conjunctions;
- when words or phrases are segregated by quotation marks or italics; and
- after short introductory terms like *in fact, for example*, and *nevertheless*.

Appropriately deleting commas permits sentences to flow smoothly. If editors replace them, so be it. The dos and don'ts of comma usage are detailed by Elliott (1997) and Alward and Alward (2000).

Semicolons

Semicolons separate closely related but independent thoughts. They indicate you are coming to a distinct—but related—thought, so slow down but don't stop. They signal pauses that are inter-mediate between commas and periods. Some effective uses of semicolons are:

- to break up strings of words or phrases which already contain punctuation;
- to join two sentences or independent clauses that are too closely related to be separated; and
- to achieve brevity by reducing four to five sentences with the same subject and verb to a single compound sentence with long but similarly constructed phrases that themselves contain multiple punctuation marks.

That last technique is effective in introductions, abstracts, or summaries where details are elucidated elsewhere. The sentence: *"This study indicated that deep rivers run slowly, except in times of flooding; shallow rivers run rapidly, particularly in rocky terrains; and creek velocities depend on bottom conditions, bank curvatures, adjacent foliage, and slopes."* resulted from the combination of five sentences and saved many words.

Semicolons are also used within sentences when a second thought is introduced by conjunctive adverbs like *consequently, even so, for example, furthermore*, or *on the other hand*. These forms are usually followed by commas when they begin sentences.

People differ on semicolon usage, so if editors suggest changes, go along with them.

Colons

Colons are specialized marks that signal more distinct pauses than semicolons. They punctuate sentences to gain emphasis achieved verbally by distinct pauses and they alert readers that important things like explanations, examples, quotations, or lists will follow. They are used: to precede bulleted lists, long explanations, series of words, and lengthy quotations; to link independent clauses when the second amplifies or explains the first; to express proportions and ratios; to separate publishers' addresses from their names in reference lists; and to follow greetings in emails, business letters, and formal correspondence.

Apostrophes

Apostrophes, the small comma-like marks raised above and after letters, are used to show where letters have been omitted in contractions and to indicate possession.

Contractions are shortened words created by replacing omitted letters with an apostrophe. Examples of contractions are: *don't, aren't, haven't*, and *it's*. The contraction *it*'s confuses many. An apostrophe before an *s* usually indicates possession (*Bill's hat*). *It's* is a contraction meaning *it is*, but *its* (without an apostrophe) represents the possessive form of *it*. To avoid confusion, spell out *it is* completely.

Using apostrophes to indicate possession is complicated. Nouns ending in letters other than *s* are made possessive by adding an *apostrophe s* (*John's* tie). Nouns ending in *s* are made possessive by placing an apostrophe after the *s* (*arthritis'* impact). Apostrophes also indicate possession by placement following initialisms and acronyms ending with the letter *s* (FAS' policies and SARS' symptoms).

The rules for using apostrophes to indicate plurality are not simple. Most nouns are made plural by adding the letter *s* (as in *instruments*). This includes the plural forms of years (the *1990s* and numbers (the *200s*). However, the plurals of letters and symbols are indicated with an *apostrophe s*.

The many exceptions regarding apostrophes are outlined by Alward and Alward (2000) and Alred et al. (2003). If still confused, follow your gut reactions but don't argue with editors unless their changes obscure your meaning.

Parentheses

Parentheses are curved vertical lines that set off incidental explanations, comments, modifiers, definitions, clarifications, or other additions without which sentences would still be complete (Alred et al. 2003). They are more emphatic than commas and less emphatic than dashes. No spaces are used between parentheses and the enclosed material.

Parenthetical expressions depart slightly from the theme of discussions. References and footnotes to acknowledge the sources of information or quotations are usually enclosed in parentheses as are cross-references like (*see glossary*). Parenthetical material does not affect the punctuation of sentences. Avoid temptations to insert distracting digressions or punctuations in parentheses.

Type Styles or Typefaces/Bold, Italic, and Underline

Italics—*vertically slanted letters*—emphasize words, titles, figures, or explanatory phrases. Italics are technically typestyles or typefaces rather than punctuation marks, but like parentheses and dashes, they are used to emphasize and set apart words so their meanings become obvious to readers. Italics identify explanatory words, letters, figures, and the titles of books, periodicals, and programs to make their role obvious. In Microsoft Word, they are accessed on the Italic icon—labeled as *I*—on the formatting toolbar, with their typestyle colleagues Bold (**B**) and Underline (<u>U</u>). The many uses of italics are detailed by Alred et al. (2003).

Boldface type is darker and thicker than surrounding words and is used to set aside letters, words, and numbers in headings, glossaries, formulas, and lists, or to achieve emphasis in many contexts. Except for headings, bold type is best used sparingly.

<u>Underlining</u> type styles also are used sparingly to set aside words or headings to create focus and emphasis.

Dashes and Hyphens

Dashes (—) and hyphens (-) are punctuations that mark pauses or indicate breaks in sentence continuity. There are two sizes of dashes. Both are longer than hyphens. Em dashes (—), also called long dashes, are as long as the letter *M;* and en dashes (–), also called short dashes, are as long as the letter *N*. Both are used to indicate hesitation, create breaks, highlight information, signal omissions, or identify words or phrases that need to be set off or emphasized. Their tendency to emphasize provides a subtle difference from parentheses, which often de-emphasize enclosed material (Elliott 1997). Dashes also introduce explanations and summaries. Usually, em dashes are not set off by spaces. Material enclosed within dashes should contain minimal punctuation.

Dashes enhance clarity. They can be used individually toward the end of sentences or in pairs that permit the reader to grasp the insertion and return to the sentence with minimal interruption. The use of dashes instead of commas and parentheses is increasing. They can be distracting, so limit their use to significant insertions.

Dashes differ from hyphens. Unlike hyphens, dashes usually are absent on many computer keyboards and are accessed via alternative methods (Weverka 2001). In Microsoft Word, em dashes are typed by striking Alt+Ctrl and the minus sign on the numeric keyboard, and en dashes are typed by striking Ctrl and the minus sign on the numeric keyboard. When two consecutive hyphens are typed without spaces between them, some computer programs and typesetters automatically insert dashes. Applications for the several sized dashes are detailed by Sabin (2001). If in doubt, let editors choose the most appropriate dash.

Hyphens are shorter than dashes. They are inserted between two related words to form a compound unit, such as *virologist-immunologist*. They link compound words, divide words that must be broken at the end of a line, space digits (as in telephone numbers), and join modifiers, such as *30-minute visit*. When inserting hyphens, be sure no spaces precede or follow them.

Quotation Marks

Quotation marks indicate that someone else's words are used. In Standard American Written English, two quotation marks enclose initial quotes and single quotation marks surround quotations within quoted materials. British writers switch that sequence. According to Elliott (1997), when using direct quotations it is advisable to:

- begin quotations with a capital letter;
- place closing periods and commas inside the final quotation marks;

- place colons and semicolons outside of closing quotation marks; and
- place closing quotation marks after the very last word when quotations involve more than one paragraph.

Long quotations can be set in block format with smaller type without indented paragraphs or quotation marks. In this mode, author information and citations follow and are preceded by dashes.

Exclamation Points

Exclamation points issue urgent commands or call for special attention. They are rare in technical documents and should be used with caution.

Question Marks

Question marks follow direct questions or indicate doubt. They can be omitted if a sentence itself indicates question or doubt. For example: the sentence "*We conducted this experiment to determine if the vaccine produces solid immunity.*" requires no question mark.

Ellipses or Ellipsis Points

Ellipses or ellipsis points (...) suggest that words have been intentionally omitted from quotes, or the deletion of words whose meanings are apparent from the context. Six consecutive periods indicate entire paragraphs have been omitted from quotations. Avoid ellipsis points in scientific writing because deleting words can decrease precision and raise questions.

Brackets and Braces

Brackets [ ], sometimes called square brackets, are square parentheses. Use them to:
- insert parenthetical information into sentences, words, or clauses that are already in parentheses;
- enclose explanatory words within quotations; and
- insert clarifying material that is not a logical part of a sentence.

Bracketed questions or suggestions are often inserted electronically into manuscripts by editors to clarify material without disturbing the original verbiage. These must be addressed and deleted before printing. If they survive, bracketed editorial comments will confuse readers. Often when parenthetic material needs further explanation, it is best to rewrite the information into two clear sentences. By British custom, the marks called parentheses in American English are called brackets and American brackets are called square brackets (University of Chicago Press 2003).

Braces { }, sometimes called curly braces, are used to indicate relationship between words and create further separation of parenthetic and bracketed material. They are best avoided unless used in formulas.

Bullets

Word processing programs have increased the use of bullets to emphasize items in lists. Bullets reduce verbiage, simplify reading, and help readers identify points of importance. While using fewer words, bullets usually use more page space because of their separation. They are useful when drafting talks and in PowerPoint presentations (see chapter 2). Bullets have definite uses in scientific writings, particularly in abstracting speeches and summarizing observations or conclusions which could be mired in narrative. Some journals won't publish lists in bullet form.

If bullets contain complete sentences, they should be short and may conclude with a period. There are varying opinions about punctuating bulleted material. Bullet lists are readable if the first word of each bullet is capitalized and they lack further punctuation. Some editors prefer to use lower case letters in the first word of each bullet, add semicolons after each bulleted statement, and place periods after the last word in the list. This permits the entire list to be considered a complete sentence.

While numbered lists imply that order is important, bullets present items in a non-numerical sequence. Nonetheless, bulleted lists need careful placement to achieve smooth flow. Lists in bullet form are preceded by an explanation of their relationship to the narrative. They should contain only related, comparable, and unranked items with identical structures.

Following the last bullet, skip a space. Then—without indenting—add a brief concluding sentence that summarizes the bullets and introduces the material that follows. Most computer programs offer multiple bullet styles.

Excessive use of bullets can detract from their effectiveness as indicators of importance. PowerPoint presentations composed entirely of bullets sometimes fail to maintain audience attention unless they are accompanied by stimulating narratives or illustrations. Try to limit each bulleted list to five items unless they comprise short and easily skimmed passages.

Achieving Goals with Punctuation

One function of punctuation is to signal speed and emphasis and transform written words into understandable messages. Therefore when you want readers to:

- stop momentarily and prepare for a changing message, insert a period;
- pause briefly because more is coming, insert a comma; and
- prepare for a list of similar items, insert a colon.

Punctuation also alerts readers about upcoming changes, so when you want readers to:

- expect lists of further punctuated words, phrases, or clauses, insert semicolons;
- recognize a brief clarification or definition, enclose it in commas or parentheses or set it off with dashes;
- realize someone else's words or thoughts are being used, insert them in quotation marks or cite a reference;
- navigate lists of comma-separated words, put commas before concluding conjunctions; and
- clearly perceive possession, add *apostrophe s* to most words; if the words end in *s*, add *es* or an *apostrophe*.

Reviewers, editors, proofreaders, and readers love to nit-pick punctuation. It is best not to argue with them. A succinct summary of punctuation can be found in the *Oxford Essential Guide to Writing* (Kane 2002).

CHAPTER 12

GRAMMAR

Introduction

Grammar governs the form and placement of words. It is best mastered when working on documents. Grammatical rules can be confusing. Dictionaries and grammar books such as those by Random House Publishing Company (2003) and Rozakis (2003) can help master the needed essentials. Grammar and word order improve during revision, editing, and proofreading.

As in sports and music, writing ability grows with practice and repetition. We can upgrade our writing potential with improved grammar but we don't need to be grammarians to write well. The first pages of this chapter review grammatical guidelines and computer grammar checker techniques. The rest of the chapter has definitions and details for use when questions arise.

Basic Grammatical Guidelines

Grammatical rules are flexible and continually changing. They are best learned while writing.

We can't predict what changes in grammatical practices will occur in the future. We can, however, recognize things we <u>must</u> do to meet minimal standards, things we <u>should</u> do to produce satisfactory documents, and things we <u>may</u> do to express our individuality. To use acceptable grammar, we <u>must</u>:

- begin every declarative sentence with a capital letter and end it with a period;
- be sure each sentence is complete and limited to a single thought;
- create sentences in subject-verb-object order;
- be sure subjects and predicates agree in number; and
- cleanse writings of double negatives, meaningless modifiers, and sentence faults.

In pursuit of writing excellence we <u>should</u>:
- write mainly in the active voice;
- make each paragraph begin with an introductory sentence, follow with detailing sentences, and close by introducing the next paragraph;
- cleanse writings of nonessential adjectives and adverbs;
- seek consistency in headings, fonts, references, and citations; and
- know the names and uses of the parts of speech.

To add variety and remain in compliance with reader's expectations, we occasionally <u>may</u>:
- split infinitives if it adds to clarity and brevity;
- invoke redundancy to emphasize critical points or summarize manuscripts or chapters; and
- use the passive voice to protect personal privacy or confidentiality and when performers of action are truly unknown.

Grammar can be improved by:
- keeping subjects and verbs as close together as possible;
- being sure the subject of each sentence is apparent;
- having subjects correspond with their modifiers in number;
- placing modifiers directly before the words they modify; and
- making it clear which noun every pronoun replaces.

Just as knowledge grows with comprehension of the concepts and vocabulary of a discipline, writing grows with improving grammar. Great writers use grammar books and dictionaries when they revise and edit. When this becomes a habit, their knowledge increases, their writing talent expands, and their understanding of grammar deepens.

As you consult references like *The Handbook for Technical Writing* (Alred et al. 2003), *The Gregg Reference Manual* (Sabin 2001), *The Chicago Manual of Style,* 15th Edition (University of Chicago Press 2003), and *Scientific Style and Format* (Council of

Science Editors 2006), you'll find that grammarians disagree. This means you too have flexibility.

Athletes learn the rules on the playing field but they can't predict or debate referees' decisions. Authors learn the rules by writing and not arguing with editors.

Editing, revision, and proofreading provide ample opportunity to work on grammar. After reading the opening pages of this chapter, save reading its details until you need them.

The ingredients of good writing are words, clauses, phrases, sentences, and paragraphs. The constantly changing rules help us combine them into excellent documents by use of careful:

- word selection (choice of understandable words);
- word order (syntax);
- word placement (sentence construction); and
- word compatibility (agreement of subjects with verbs, and pronouns with nouns).

The following grammatical components should gradually be mastered on the road to writing excellence.

Grammatical Components

The components of grammar can be overwhelming if memorized in the abstract. They become meaningful when their definitions are learned while improving documents. When the terms that follow are encountered in editors' comments, reading about them will expand your writing arsenal.

Words and Word Selection

Writing starts with choice and placement of words. Strive to select words that are: simple, short, and easily understood; appropriate to convey intended meanings; and in the best possible order.

Simple, Understandable Words

The choice of simple, easily understandable words permits readers to grasp the message. Choose words to express ideas, not to impress readers. Sophisticated or lengthy words display our intelligence but make readers stumble. Replace uncommon words with simple terms that don't require rereading to figure out what they mean. Reader-friendly word selection avoids gobbledygook.

Syntax: Word Order and Arrangement

Syntax, the selection and arrangement of words, phrases, and clauses to form sentences, comes into play once you decide what a sentence is to say. Three components of effective word placement are subject-verb-object order, the active voice, and modifier positioning.

Subject-Verb-Object Order

Words flow naturally into place if sentences start with a subject that is followed closely by a verb (predicate) and then by the recipient of an action (object). This subject-verb-object order is most effective when used in the active voice.

The Active Voice

In the active voice, subjects are described or perform actions. It is more direct than the passive voice (see chapter 5). The passive voice has the subject receiving—rather than performing—action and creates long and complicated sentences.

Modifier Positioning

The placement of modifying words and passages is one key to word order. During revision and editing, see that modifiers are meaningful and located so it is obvious which words they modify. Then search for dangling, misplaced, and meaningless modifiers.

Dangling modifiers hang out as loners and make readers wonder what they modify. Misplaced modifiers accidentally modify the wrong word. Meaningless modifiers are usually nonspecific terms like *crucial, important,* or *great,* that add little to the message. Meaningless modifiers are identified by asking *"Compared to what?"* Unless that question is answered, they are best deleted.

Good word placement requires familiarity with the names and functions of the parts of speech.

The Parts of Speech

Words are classified as parts of speech by function and placement. The parts of speech are nouns, pronouns, adjectives, verbs, adverbs, conjunctions, prepositions, and interjections. They are used as follows:

- nouns name persons, places, or things;

- pronouns replace nouns;
- adjectives describe or modify nouns, pronouns, or other adjectives;
- articles, a class of adjectives (*a*, *an*, and *the*), modify and limit nouns;
- verbs express action, being, or assertions;
- adverbs describe or modify verbs;
- conjunctions connect items;
- prepositions introduce phrases; and
- interjections intercede to create emphasis.

Nouns, pronouns, adjectives, verbs, and adverbs are the work-horses of writing excellence. Conjunctions and prepositions are the harnesses and hitches that keep the team moving together. Interjections are nonessential decorations added to attract attention.

Authors are the drivers. They direct movement and pace. The smoothness of the ride increases when the driver understands the individual behavior of each workhorse and the functions of the harmonizing apparatus that keep them operating in synchrony. The names of the parts of speech appear repeatedly in writing guidelines. Understanding them helps us convey our meanings.

Many words serve as several parts of speech. For example: the words *test*, *work*, and *transplant* function as both verbs and nouns. Dictionaries accompany words with bold abbreviations indicating a part of speech for each definition. Most dictionaries use: **adj.** for adjective, **adv.** for adverb, **conj.** for conjunction, **interj.** for interjection, **n.** for noun, **pron**. for pronoun, **prep**. for preposition, and **v.** for verb. These abbreviations may vary.

Using Computer Grammar Checkers

Computer grammar checkers usually run simultaneously with spell-checkers. In Microsoft Word they are activated by clicking the spell-checker icon on the toolbar. Many writers leave the grammar and spell-checkers on all the time. If the grammar checker underlines a passage in green, it means something doesn't look right. It may be a double space, a missing apostrophe, or other defects. Grammatical savvy is enhanced by seeking ways to erase the green underline before selecting the spelling-grammar icon for an explanation.

Grammar Details

The rest of this chapter contains grammatical details. Use it as a reference when questions arise or read it in small doses.

The Parts of Speech Detailed

If writers understand the parts of speech—which are best learned individually while addressing questions—and rely on the computer grammar checker, they can get along without memorizing the following details or the definitions in the glossary.

Nouns

Nouns name what we write about. They often constitute subjects, start sentences, and govern sentence flow. Nouns name persons, places, or things. They also designate or identify entities, qualities, states of being, actions, concepts, or ideas. They are modified by adjectives and can be replaced by pronouns. Nouns can also serve as adjectives or adverbs. They appear in every sentence and have many uses. They can function as subjects, predicates, complements, direct and indirect objects, or objects of prepositions. Those terms are addressed later.

Nouns are designated by number (singular or plural); gender (feminine, masculine, or neuter); and person (first, second, or third). Also, they are categorized by case (nominative, objective, or possessive) depending on their role in sentences. Correct selection of nouns and noun forms comes unconsciously to English speakers. Noun usage is challenging if you grew up where English was not the first language or where good English was in short supply.

Noun number causes relatively few problems because singular and plural forms are usually evident. The gender of animals, plants, and people (female, male, or neuter) is usually clear and often irrelevant.

The person of nouns indicates their role in sentences by expressing the writer's relationship to the reader and the content. First person nouns (*I* and *we*) represent the person(s) writing, second person nouns (*you*) represent the reader(s), and third person nouns represent the people, animals, or things written about. Nouns themselves rarely change form to adjust for person but they dictate the person of the pronouns (discussed below) that replace them.

The case of nouns is dictated by their use in sentences. Nouns used as subjects take the nominative case. Nouns used as objects are recipients of actions or descriptions and take the objective case. Nouns change form only in the possessive case. Singular nouns are made to show ownership by ending in an *apostrophe s*. The possessive of plural nouns that already end in *s* or *es* is formed by adding an apostrophe (*teachers'* obligations or *vaccines'* effectiveness).

The many types of nouns include:

- Common nouns identify general groups of persons, places, or things, and are not capitalized unless they start a sentence. Examples are *companies* and *agencies*.
- Proper nouns name specific companies (*Dow Chemical*), agencies (*FDA*), or institutions (*Stanford University*), and specific people, cities, states, countries, or organizations, and are capitalized regardless of their location in sentences.
- Collective nouns refer, without specifically naming their members, to groups (such as the *faculty* or the *agency*). They may or may not be capitalized, but are considered singular and take singular verbs, pronouns, and adjectives. Collective nouns can also be used as singular adjectives as in *faculty* meeting or *agency* policy.

Writers often develop concepts using abstract nouns such as *livestock*, *feed*, or *illness*, and subsequently detail facts with concrete nouns (like *cattle*, *corn*, and *influenza*).

Steinmann and Keller (1999) and Hopper et al. (1990) use the term *nominal* to describe nouns that are derived from verbs. For example: *cauterization* and *pasteurization* come from the verbs *cauterize* and *pasteurize*. Nominals also include clauses and phrases that assume noun-like roles to describe or name people, objects, or places. Grammatical usage of the terms *nominal* and *nominalization* causes confusion because the word *nominal* has many meanings, including *little* and *trifling*.

Noun Clauses and Noun Phrases

Nouns are obvious as single words naming persons, places, or things, and when serving as the subjects of sentences. Phrases and clauses can function as nouns and serve as subjects, objects of verbs or prepositions, or predicate nominatives (words or phrases connected to nouns by linking verbs that explain or rename them).

Noun clauses and noun phrases are difficult to recognize because they are loaded with non-noun words. This concealment blurs the fact that they must follow the rules imposed on one-word nouns. For example: noun clauses and noun phrases often begin with relative pronouns like *who, whom, whose, which, what,* and *that.* An example of a noun phrase is *sharpening knives,* which serves as the subject of the sentence "*Sharpening knives* is a daily ritual for slaughterhouse workers."

Successful Use of Nouns

Seek the correct meaning and accurate spelling of nouns. Eight secrets to success in noun usage are:

First, seek correct plural forms. Nouns usually add *s* unless they end in *s, sh, c, ch, i,* or *z* and require an *es* ending. Some nouns have irregular plural forms, such as calf/calves, bacterium/bacteria, mouse/mice, nucleus/nuclei, serum/sera, and vertebra/vertebrae. Other nouns, like series, sheep, salmon, deer, swine, and moose are identical in both singular and plural. When in doubt consult a dictionary.

Second, be sure noun subjects agree with their verbs in person (first, second, or third) and in number (singular or plural). This is challenging when phases or clauses containing plural nouns are placed between singular noun subjects and their verbs. The sentence "This disease (*singular subject*), which persists in most states and has many fatal ramifications (*plural modifiers*), is (*singular verb*) highly contagious" tempts us to overlook the singular subject (*disease*) and use a plural verb (*are*).

Third, assure that noun antecedents agree in case with the pronouns replacing them. Use the nominative case for subjects, the objective case for objects, and the possessive case for ownership. Also assure pronoun-antecedent concurrence in number (singular or plural).

Fourth, replace abstract nouns with concrete nouns unless they are in introductory sentences that will be followed by detailed descriptions using specific terms.

Fifth, use care in expressing possession by nouns. Usually the possessive form is indicated by adding *apostrophe s* (*'s*) unless the noun ends in *s* and thus requires only the addition of an *apostrophe* (*s'*). Some books distinguish between possessive forms of singular and plural nouns. The rule is: for singular nouns, add *'s* even if the noun ends in *s* (like *boss's*), and for plurals ending in *s*, add an apostrophe only. Possession can also be indicated by possessive pronouns such as *my, their,* or *our,* which don't require an apostrophe.

Sixth, be cautious with nouns derived from verbs. Words like *examination, vaccination,* and *sterilization* contribute to wordiness because they need helping verbs to support them. *We vaccinated the dogs* or *The dogs were vaccinated* are clearer and shorter than sentences like *The dogs were subjected to vaccination* or *Vaccination was carried out on the dogs.*

Seventh, be sure strings of nouns used as modifiers are similar in case and number and are separated by commas. If you have written *This additive is safe for horses, goat, bovine, or pig's feed"* change it to *This additive is safe for horse, goat, cattle, or pig feed.* This change creates parallel structure and is easier to read.

Eighth, be sure appositives, which are nouns or noun phrases that clarify the meaning of nouns, are closely aligned with the noun they modify and are set off by commas. The italicized words comprise the appositive in this sentence: Serology, *the least expensive test,* was effective in obtaining a diagnosis.

The above advisories expedite drafting, revision, and editing. Nouns abound and new ones are continually emerging. When nouns are replaced by pronouns, they become antecedents that dictate the characteristics of the pronouns that replace them.

Pronouns

Pronoun means "for a noun" in Latin. Pronouns substitute for nouns. They expedite reading by replacing nouns with shorter words and reducing repetition. The nouns pronouns replace are their antecedents, i.e., "their predecessors." To avoid reader confusion, writers must assure that pronouns agree with their antecedents in person (first, second, or third), number (singular or plural), and gender (feminine, masculine, or neuter). This requires recognition of the person, number, and gender of the nouns.

Pronouns have multiple forms and uses. They are classified as personal pronouns, demonstrative pronouns, indefinite pronouns, possessive pronouns, relative pronouns, interrogative pronouns, and reflexive pronouns. Some pronouns serve in several categories. Detailed knowledge of this classification is not essential to writing excellence but helps when questions arise.

Personal Pronouns

Personal pronouns are small words that replace nouns and accurately mimic them regarding number and gender. They also specify the relationship of their antecedents (and attached verbs) to sentences by grammatical categories called person.

First person pronouns indicate the writer(s) as *I* (singular), *we* (plural), and *my* or *mine* and *our* or *ours* (possessive). Second person pronouns refer to individuals that are written to, as *you*, (singular and plural) and *your* or *yours* (possessive). Third person pronouns indicate the persons or things that are written about, as *he*, *she*, or *it*.

First person personal pronouns (*I*, *we*, *us*, and *ours*) are usually avoided in technical writing but this is changing because they are direct and simplify reading. Without them, writing slips into the passive voice and becomes verbose (see chapter 5). Second person personal pronouns address the reader (*you*, *your*, and *yours*). They are omitted when making suggestions, issuing instructions, or outlining protocols. For example: the pronoun *you* can be deleted in statements like *Incubate the culture at 36 degrees Centigrade for 72 hours.* Third person personal pronouns such as *he, him, his*; *she, her, hers*; and *they, them, their*, or *theirs* refer to people written about and require clearly identified antecedents. However, the identities of persons identified as *I* or *me* (first person) and *you* (second person) are self-evident, so they are used without antecedents.

Consistency of person is essential in sentences, desired in paragraphs, but not required throughout documents if transitions from description (third person) to instructions or advice (second person) are evident to readers. Switching the person of subjects is permissible but requires that pronouns match their antecedents in case, number, and gender.

Demonstrative Pronouns

Demonstrative pronouns (*this, that, these,* and *those*) replace nouns and single them out for further attention. The demonstrative pronoun *these* is used in the sentence: Seven endangered species are present in the park and *these* require protection. Demonstrative pronouns are often correctly used as subjects of sentences when the actual subject is named in the preceding sentence. For example: The symptoms (plural noun subject) of influenza are multiple. *These* (demonstrative pronoun) are similar in many respiratory diseases.

Indefinite Pronouns

Indefinite pronouns refer to groups, classes, or collections of people or things and sometimes to specific individuals or items. They comprise the largest class of pronouns and can cause confusion. They are called *indefinite* because they can be used without antecedents and can refer to remotely located nouns. Some words serving as indefinite pronouns are:

- one, anyone, everyone, each one, each, no one, none, someone, everyone;
- anybody, everybody, somebody, nobody;
- anything, everything, nothing, something;
- all, any, each, either, few, neither, several; and
- another, other, some, such, many, neither.

Indefinite pronouns can also be used as nouns and serve as subjects. The nouns that indefinite pronouns replace must be understood from previous sentences, or clarified in subsequent sentences. They can cause confusion when these conditions are not met because other nouns can separate indefinite pronouns from their antecedents. Unless immediately followed by clarifying modifiers, indefinite pronouns should be used cautiously. They are best replaced by specific terms.

Possessive Pronouns

Possessive pronouns show ownership or possession by their antecedents. Some possessive pronouns are: *my, mine, your, yours, hers, his, its, our, ours, their,* and *theirs.* They lack the apostrophe their noun counterparts use to express possession. When using possessive pronouns, be sure it is clear which nouns they refer to.

Relative Pronouns

Relative pronouns also replace nouns. They link or convert words, usually by introducing subordinate clauses like "The boy *who was exposed to measles* never got sick." They also convert or relate independent clauses to main clauses. Some relative pronouns are the various forms of *who* (*who, whom,* and *whose*), *which,* and *that.*

The word *who* replaces a noun naming people; the pronoun *which* replaces nouns naming things, animals, or other living things—but not people; and the relative pronoun *that* is flexible. It can be used to replace nouns that name animals and nonliving objects—and less commonly, people. As an example: "The rancher *who* owned the cattle, *which* died, also used the fertilizer *that* was incriminated." That fifteen-word sentence is clear and grammatically correct but could be correctly written in ten words without relative pronouns, as: "The owner of the dead cattle used the incriminated fertilizer." Thus, complicated pronoun rules can sometimes be avoided by rewriting sentences. Editors have differing views on relative pronouns and it is best to agree with them. Some guidelines for relative pronouns are:

- use *who* for people (the lady *who*...);
- use *which* for nonhuman creatures (the wildlife which...); and
- if in doubt, use *that* because it is acceptable for almost everything except people.

The pronoun *who* can be confusing. It serves in the nominative case. Its objective case is *whom* and its possessive case is *whose.* Whom (objective case) implies that someone received the effect of an action. To decide whether to use *who* or *whom* as the object of a verb or preposition, mentally rephrase sentences and determine if the noun in question can be replaced by *he* or *she* or by *him or her.* If *he* or *she* fits, use *who.* When the word *him* or *her* sounds better, select *whom.* Some grammarians say the distinctions between the words *who* and *whom* are irrelevant.

Interrogative Pronouns

Interrogative pronouns such as *who, which,* and *what* often begin sentences that ask questions and close with a question mark. Their antecedents should be clearly evident.

Reflexive Pronouns

Reflexive pronouns usually end in *self* (*myself, yourself,* and *themselves*). They refer to noun subjects indicating that they act upon themselves, as in "I hurt *myself*" or "They hurt *themselves.*"

They must replace a noun in the <u>same</u> sentence. Thus, the sentences "*Myself* sent the letter" and "The letter was sent to *myself*" are unacceptable in writing. Reflexive pronouns can cause problems with regional dialects that conversationally misuse them as subjects or objects when the noun antecedents (*I*, *you*, or *we*) are missing.

Pronoun Antecedent Agreement

Using small pronouns to replace large nouns or nouns with multiple modifiers reduces wordiness, but requires careful pronoun positioning so readers can identify the nouns they replace. During revisions, examine all pronouns to assure they are close to their antecedents and agree with them in number and gender.

Pronoun number identifies their antecedents as singular or plural. Many pronouns take different forms in the singular and plural. We intuitively shift from *I* to *we;* and from *he*, *she*, or *it* to *they* to indicate plurality. Some pronouns, such as *each, everybody, either, one*, and others are generally considered singular (Maas 2000). Others including *both, few, many*, and *several* replace plural nouns. Still other pronouns like *all, any, some*, and *none* can replace either singular or plural nouns. Consistency in number between pronouns and their antecedents is guided by complex rules. The number of a pronoun is often evident, but where questions exist, let your intuition prevail unless editors object. Dictionaries provide little help in this area.

Gender agreement between pronouns and antecedents is undergoing changes due to movement toward gender neutrality. If a pronoun replaces the name of a specific individual, it is appropriate to use the words *he, she, his*, or *hers*, but if you are choosing a pronoun to replace gender-neutral nouns like *farmer* or *doctor*, you must choose between *he* and *she*. You can avoid this dilemma if you initially choose a plural form (*farmers* or *doctors*) and then refer to them as *they*. Consult *The Gregg Reference Manual* (Sabin 2001) to perfect pronoun-antecedent agreement.

Successful Use of Pronouns

Successful pronoun use requires each pronoun's antecedent to be readily apparent. This is achieved by placing pronouns close to the nouns they replace and having no other nouns between them. Fundamental concepts of pronoun use are:

- The case of pronouns is determined by their function in sentences, is dictated by the nouns they replace, and can be of the nominative case, objective case, or possessive case.
- The person of pronouns is first (*I, we*) if they indicate the person(s) writing; second (*you*) if they indicate the person(s) written to; and third (*he, she, it,* or *they*) if they indicate the person(s) or the things written about.
- The number of pronouns can be singular (*he, his, she, hers, it, its*) or plural (*they, theirs*).
- The gender of pronouns can be feminine (*she, her, hers*); masculine (*he, him, his*); or neuter (*it, they,* or *theirs*).
- When pronouns serve as subjects of sentences, it should be clear which nouns they replace and they should agree in number.

Pronouns are little words whose judicious use can expedite clear writing and easy reading. An excellent summary of pronoun problems appears in *The Only Grammar Book You'll Ever Need* (Thurman 2003).

Adjectives

Adjectives modify or limit nouns, pronouns, and other adjectives. They specify size (the *big* dog*)*, shape (the *round* ball), color (the *red* barn), or number (the *10* girls). They also indicate which one (*that* dog or *this* crop); what kind (*multivalent* vaccine); or how many (*five* balls, *several* crops, or *few* apples). Nouns and pronouns can function as adjectives, as in the *farm* truck or the *antibiotic* ointment.

The most common adjectives are the articles *(a, an,* and *the)*. Some texts address them as separate parts of speech called determiners (Maas 2000) or noun markers (Random House Publishing Company 2003) because they establish usage of nouns without modifying them. The indefinite articles (*a* and *the*) modify singular nouns indicating there is only one. The article *a* precedes singular words beginning with a consonant sound while the article *an* precedes words starting with vowel sounds. However, these choices depend on sound and not the actual first letter of the modified word ("a historic moment" rather than "an historic moment"). Vowel sounds are usually the most prominent sounds in a syllable and resemble the common pronunciations of *a, e, i, o, u,* and sometimes *y. The* is the only definite article. It identifies specific things.

Proper adjectives are derived from proper nouns. Examples are *Washingtonian* attitudes, *Christmas* parties, and *African* ecosystems. In addition, there are compound adjectives like *soft-shell* crab and *tissue culture* cells. Certain words, called absolute adjectives, cannot be further modified (Thurman 2003). It is awkward to say something is *very* dead, *quite* unique, or *rather* perfect because the adjectives *dead, unique,* and *perfect* are considered final.

When expressing degrees of comparison, adjectives (as well as adverbs) assume one of three forms. The positive form appears in dictionaries and is the basic mode or degree (*fast* or *good*). The comparative form (*faster* or *better*) compares two things, and the superlative or speculative form (*fastest* or *best*) is used to compare three or more things. These forms of *good* comprise an adage applicable to writing. It says "*Good, better, best; never let it rest; till your good is better; and your better best.*"

When editing, check adjectives for agreement with the words they modify with respect to gender and number. Adjectives can be overused. Wordiness can be reduced by deleting nonessential adjectives. Prime candidates for deletion are *very, slightly, almost, few, crucial,* and *important.* In selecting those to be eliminated, read sentences aloud to see if dropping adjectives changes the meaning.

Successful Use of Adjectives

Here are some recipes for effective use of adjectives.

First, when editing or revising, carefully scrutinize all adjectives and ask if they really add clarity, precision, and accuracy or if they merely occupy space. Delete those that are unessential, meaningless, or prejudicial.

Second, when using possessive adjectives to show ownership, place them before the nouns they modify.

Third, when using adjectives to modify subjects, have them agree in gender and number.

Fourth, when two adjectives collectively modify the same noun, they may be hyphenated to suggest merger as in *real-time* PCR or *electronic-mail* messages. When one adjective modifies another adjective, as in *very rapid,* hyphens are usually not used.

Fifth, when using adjectives in a series, insert commas between those that modify nouns (a *tall, lanky, blond* boy) but not between those modifying other adjectives (a *very fast* race).

Sixth, when using compound adjectives, test comma use by replacing commas with the word *and*; if *and* doesn't sound right, the comma can usually be omitted.

Seventh, when proper nouns are used as adjectives to modify other nouns, be sure they are capitalized, as in the *Paris* meeting or the *October* hurricane.

Eighth, when possible, choose precise words or numbers rather than vague terms like *many, few*, or *numerous*.

These suggestions, along with careful deletion of unneeded adjectives and other excessive or meaningless modifiers (see chapter 5), will increase brevity and clarity.

Verbs

Verbs state action or being. There are regular verbs, irregular verbs, transitive verbs, and intransitive verbs.

Regular verbs retain their basic root but add *ed* or *d* to form the past tense and *ing* to assume their participle forms. Thus, *I fix, I fixed, I have fixed*, and *I have been fixing* constitute the conjugation of a regular verb. Most verbs are regular and present few problems when they change tense.

Irregular verbs change form in the past tense, so their past tense must be memorized or found in a dictionary. The principal parts of irregular verbs are the forms they take in different time settings, i.e., the present tense, past tense, future tense, present participle, and past participle. Most dictionaries list the principal parts of irregular verbs (bleed, bled, and bleeding) to indicate present tense (we bleed), past tense (we bled), and past participle (we were bleeding). Some irregular verbs are:

- begin (present tense), began (past tense), and begun (past participle);
- eat (present tense), ate (past tense), and eaten (past participle);
- freeze (present tense), froze (past tense), and frozen (past participle); and
- write (present tense), wrote (past tense), and written (past participle).

These and many other commonly used irregular verbs are memorized and come naturally to English speakers. When doubt exists, consult a dictionary.

Transitive Verbs

Transitive verbs perform actions and require an object. They indicate that the subject of a sentence has done something to someone or something. The recipient or direct object of the action must follow the verb to make it clear what happened and to complete the sentence, as in *"He hit the ball."* The direct object is the immediate or straightforward receiver of an action and indirect objects are secondary recipients. Thus in the baseball example, if an indirect object emerges, the sentence is amplified to say "He hit the ball (direct object) to the shortstop (indirect object)."

Intransitive Verbs

Intransitive verbs are action verbs that indicate that the subject of a sentence has done something but that no person, animal, or thing was affected and thus no object is needed. They express actions such as *run*, *walk*, and *breathe*, which are complete without objects or linking verbs. Examples are "The doctor *meditated*" or "The boy *slept*."

Linking Verbs

Remember, verbs express action or being. Verbs indicating being rather than action are called linking verbs. They connect subjects with their predicates or modifying adjectives and they usually don't take objects. Linking verbs add to the understanding of subjects by describing their characteristics or traits as in "Heart surgery *is* risky" or "He *seems* nice" without naming a recipient of actions. Most linking verbs are forms of *be* (*am*, *are*, *is*, *was*, or *were*) and words such as *appear, look*, and *become*. They are often followed by adjectives as in "The vaccine *is* (linking verb) effective (adjective)."

While action verbs are usually obvious, verbs that express being—usually forms of the word *be*—are more subtle. They take different forms in the present tense (*am*, *are*, and *is*), the past tense (*was* and *were*), and in their participle forms (*being* and *been*). These are routine for English speakers but are challenging for nonnatives and translators.

Verbals

Verbals are verbs forms that function as nouns, adjectives, or adverbs. They can include gerunds (verb forms ending in *ing*) as in "*Running* is fun," infinitives (verb forms following *to*) as in "It is fun *to run*," and participles (verb forms ending in *ing* and functioning as adjectives) as in "The *running* boy tripped."

Because they function as nouns, adjectives, or adverbs, verbals cannot comprise the only verb form in a sentence (Maas 2000). Dictionaries and grammar guides differ in their definitions and explanations of verbals and writers use them correctly without recognizing them. Therefore don't worry about them unless an editor raises the issue.

Voice

Verbs function differently in the active and passive voices. Voice is verb usage that designates subjects of sentences as doers (active voice), as in "*The dog bit the boy,*" or as receivers of actions (passive voice) as in "*The boy was bitten by the dog.*" The active voice is more direct and less wordy than the passive voice. One step in revision is to replace the passive voice with the active (see chapters 5 and 6).

Mood

Mood indicates the grammatical function of verbs and reflects the feelings of writers regarding the audience and the message. Verbs of the indicative mood state something, as in "The appendix *ruptured* during surgery." Verbs of the subjunctive mood postulate, as in "The appendix *might have ruptured* during surgery," and verbs of the imperative mood issue commands, like "*Remove* the appendix before it ruptures."

Most writing uses the indicative mood. Limited use of the subjunctive mood creates variety and enables authors to speculate or hypothesize. The imperative mood is appropriate when issuing instructions or outlining protocols.

Subject-Verb Agreement

Verbs must agree with their subjects in number and person. Singular subjects take singular verbs and plural subjects take plural verbs. This can be challenging (Steinmann and Keller 1999) when:

- sentences begin with undefined subjects or subjects that follow verb forms;
- sentences begin with noun clauses that contain both singular and plural nouns; and
- compound subjects are connected by *either-or* or *neither-nor*.

When editing, be alert for these challenges and be prepared to address them positively to assure that readers follow your reasoning.

Because verbs describe actions, procedures, situations, and states of being, they require particular attention. Some key points for verb use are:

- verbs must agree with their subjects in number (singular or plural);
- verbs must agree with their subjects in person (first, second, or third);
- simple verbs are best unless there is reason to slow reading or vary style; and
- the adverb *up* requires caution because when attached to verbs, it makes readers stumble, so replace *look up* with *seek*; *read up* with *study*; and *grind up* with *grind*.

Further cautions arise from similarly pronounced verbs with different meanings. Review this list to identify words that may trip you up:

- *Censor* means to remove objectionable material, while *censure* means to officially denounce or reprimand.
- *Precede* means or go before in time, while *proceed* means to continue after stopping.
- *Raise* is a transitive verb meaning to elevate or lift something up, while *rise* is an intransitive verb meaning to go up or get up.
- *Utilize* means to make use of, while *use* means virtually the same thing but is less stilted and more informal. Both are sometimes replaced by the nominal form (*utilization*), which requires verbal support and introduces wordiness.

Invoking these guidelines for using verbs creates brevity, clarity, and improved readability. Some grammarians consider verbs the most important part of speech, so pay special attention to them when revising and editing.

Adverbs

Adverbs modify or explain verbs, adjectives, or other adverbs. They have many uses. They qualify actions and states of being with respect to when, where, how, how often, or how much (Alred et al. 2003). Some words serve as both adverbs and adjectives and they are sometimes discussed collectively. However, adverbs usually modify verbs while adjectives usually modify nouns, pronouns, or other adjectives. Many adverbs are formed by adding the suffix *ly* to adjectives.

Conjunctive adverbs are used with a preceding comma and a following semicolon to smoothly join independent clauses into sentences (see chapter 11). Some examples of conjunctive adverbs are: *also, thus, hence, moreover,* and *nevertheless* (Thurman 2003).

Caution is needed to guard against overuse of *very, more, most,* and *best* if more specific modifiers are available. These adverbs, called intensifiers (Thurman 2003), along with others like *awfully, extremely, quite, really,* and *somewhat* can be eliminated during revision and editing (see chapters 5 and 6).

Try to have adverbs precede the verbs they modify. When that is not feasible, place them close to the words they modify. Be alert for adverbs that attempt to intensify adjectives or other adverbs without adding new information. It is best to delete *quite, really, kind of, most,* and similar nonspecific qualifiers. Carefully used adverbs add clarity.

Prepositions

Preposition means to put forward or place before. Prepositions serve mainly as opening words in phrases that modify and strengthen nouns. They do this by indicating relationships such as time (*after, before, during*) and position (*above, beyond, behind,* or *below*). By introducing prepositional phrases, they relate nouns or pronouns and their objects to other sentence parts by combining them with modifiers that add additional information.

Some one-word prepositions are: *for, with, after, before, during, until, to, into,* and *toward.* Brevity is enhanced when multi-word prepositions are replaced with single words. Thus, *in regard to* can become *regarding, in addition to* can become *plus,* and *with the exception of* can be replaced by *except*. Repeat the preposition when prepositional phrases or clauses are used in a series. For example: "We examined the pastures *for* automobile batteries, *for* insecticide containers, and *for* oil disposal sites." Remember, short or one-word prepositions in titles or headings are usually not capitalized unless they are the first word or are words the author wishes to stress.

Prepositional phrases often appear at the end of sentences. They consist of the introducing preposition, its object, and its modifiers. They are most effective when placed close to the word they modify. Some secrets for successful use of prepositions are:

- Remember, prepositions never function alone. They always require objects.
- Try to have prepositions precede their objects.
- Try to avoid repetitive or redundant prepositions such as *alongside of, off of,* and *inside of.* When these are recognized, simply delete the word *of.*
- Make pronouns serving as objects of prepositions take the objective case (*me, him, her,* or *them*).

Careful use of prepositions makes easy reading.

Conjunctions

Conjunctions are connecting words such as *and, or, but, so,* and *also*. They indicate relationships between words, phrases, and clauses. Their effective use improves brevity and clarity. If misplaced they can cause confusion. Conjunctions are classified as coordinating, correlative, and subordinating conjunctions and some (such as *but*) serve in several categories.

Coordinating Conjunctions

The coordinating conjunctions (*and, but, for, or, nor, so,* and *yet*) connect similar terms. They connect adverbs to adverbs, clauses to clauses, and phrases to phrases.

Correlative Conjunctions

Correlative conjunctions also connect similar words. They are often used in pairs to link individual thoughts or words. Examples are *either-or, neither-nor, not only-but also,* and *whether or not.* Other correlative conjunctions are *if, because, while,* and *that.*

Subordinating Conjunctions

Subordinating conjunctions (*after, as, before, since, unless, when,* and *whereas*) connect adverb or noun clauses to main clauses. As the name suggests, subordinating conjunctions indicate that one statement is less significant than another.

Use of Conjunctions

Conjunctions themselves rarely cause problems but punctuating them can be tricky. Recipes for successful use of conjunctions are:

- Use commas before coordinating conjunctions that introduce independent clauses which could stand alone.
- Remember that when two words are joined by *and*, they take plural verbs; but when two words are joined by *either-or* or *neither-nor*, they take singular verbs.
- Avoid starting sentences with conjunctions. Sentences occasionally start with *and* or *but* when emphasis is desired. Avoid this technique.

Attention to choice and placement of conjunctions will improve clarity.

Interjections

Interjections are words or phrases inserted to gain attention or express surprise or emotion. They often have little relationship to the rest of a sentence and are usually followed by exclamation points or commas. Common interjections are *oh, ah, nonsense,* and *wait.*

Interjections are most common in speech, less common in fiction, and rare in technical writing. The recipe for interjections is: Hey! Avoid interjections in publications and use them sparingly in interoffice correspondence.

A Perspective on the Parts of Speech

Study the names, characteristics, and uses of the parts of speech after you are well along the road to writing excellence. This familiarity will aid manipulation of words into clauses, phrases, and sentences; help determine the uses and limitations of words; and increase understanding of editors' comments. Save mastery of details until specific questions arise.

Clauses

A clause is a group of words with a subject and a predicate. There may be several clauses in a sentence and they can function as nouns, adjectives, or adverbs. There are independent and dependent clauses.

Independent Clauses

Independent clauses express complete thoughts. They would be sentences if punctuated with initial capitals and final periods. Every complete sentence contains at least one independent clause and many contain several.

An error called a comma splice occurs when two independent clauses are joined only by a comma. An example of a comma splice is: "The dermatologist thought it was a harmless lipoma, the surgeon suggested removal." This comma splice can be corrected by:

- inserting the conjunction *but* after *lipoma*; or
- placing a period after lipoma and starting a new sentence with *The surgeon;* or
- replacing the comma with a semicolon.

If two independent clauses are connected without punctuation they are called run-on sentences or fused sentences. The options for correcting them are:

- separating the two independent clauses into distinct sentences using periods and capitals;
- connecting the two independent clauses by semicolons and appropriate verb forms; or
- separating the two independent clauses by a comma and a coordinating conjunction such as: *and, or, but*, or *yet*.

Dependent Clauses

Dependent clauses require further information to express a complete thought. They have subjects and predicates but they lack meaning unless they are attached as modifiers to nouns or pronouns. Dependent clauses are usually introduced by subordinating conjunctions (such as *when, where, before, as, since*, and *because*) or relative pronouns (like *who, whom, which, what*, and *that*). A sentence using two dependent clauses is: "The laboratory technicians *who ran the test* were taken to the hospital *when they sickened."*

Attention to effective use of clauses makes for easy reading and is best accomplished during editing and revision. While clauses have both subjects and predicates, phrases lack one of them.

Phrases

Phrases are groups of related words with either subjects or predicates but not both. They can serve as nouns, adjectives, verbs, or adverbs, and can express complete thoughts. Correct use of phrases improves writing and their inappropriate use causes confusion. Phrases serve many purposes. They:

- modify or expand any part of a sentence when preceded or followed by punctuation;
- stand alone as colorful expressions, slogans, rules, or sayings, such as *Stars and Stripes Forever;*
- serve as nouns, adjectives, verbs, or adverbs within sentences; or
- introduce sentences from which they are separated by a comma, as *"In order to complete the agenda,* we must start on time."

The many kinds of phrases are named according to their use and function. Some are:

- noun phrases, which contain only nouns and noun modifiers, serve as subjects or objects, as in the sentence *"Rapidly expanding abscesses* surrounded the thorns in her toe."
- verb phrases, which contain verbs and helping verbs and can open questions, as in *"Are they testing for viruses* in that laboratory?"
- prepositional phrases, which behave as adjectives or adverbs to modify nouns or verbs and are introduced by prepositions that relate to or join with their noun or pronoun objects to qualify them or set them aside, as in "Milk *with high somatic cell counts* suggests udder infection."
- participial phrases, in which verbs are used as adjectives to modify nouns or pronouns, as in "The serum *selected by the committee* will be the standard."

- infinitive phrases that utilize infinitives (*to* followed by a verb and appropriate modifiers) to spell out one action part of a sentence, as in "We diluted the lye *to reduce its irritability.*"
- gerund phrases, those that use gerunds (*ing verbs*) as subjects, as in "*Testing for viruses* is challenging" or as direct objects, as in "They tried *testing for viruses.*"

Appropriate use, placement, and punctuation of phrases will enhance writing. Words, clauses, and phrases are vital to writing excellence. Sentences shape documents.

Sentences

Sentence orderliness and flow determine the readability of documents. Simple sentences convey a single thought and need only a subject and a predicate, but most sentences also contain objects and/or modifiers.

Subjects, usually nouns, appear early and indicate what sentences are about. The predicate describes the subject's activity, condition, or status. The subject usually precedes the predicate, but this order can be reversed if the sentence asks a question or is written in the passive voice.

Rework sentences repeatedly to assure that they convey a single thought and have clear meanings. Sentences should have their subjects near the beginning and near their predicates, have subjects and predicates that agree in number and person, and have pronouns that agree with their antecedents in person, number, and gender. These goals become more achievable with writing and revision experience.

Sentences are categorized by their structure, intention, and style (Alred et al. 2003). There can be *simple, compound, complex,* or *compound-complex* sentences. Simple sentences express a single thought and contain one independent clause. They can contain just a subject and a predicate but may also have objects.

Compound sentences contain two or more independent clauses connected by punctuation. They express at least two separate thoughts and have at least one independent clause. Compound-complex sentences have two or more thoughts expressed in at least two independent clauses and one dependent clause. Compound-complex sentences are usually long. They challenge readers and are best avoided. Good writing uses a mix of simple, compound, and complex sentences.

Based on their purpose, sentences also are classified as declarative, interrogative, imperative, or exclamatory. Declarative sentences make statements or convey information and dominate writing. Interrogative sentences ask questions. Imperative sentences issue commands and are used in protocols. Exclamatory sentences are used for emphasis, commands, feelings, opinions or concerns, or to suggest something is extraordinary or remarkable. They have few applications in technical writing.

Sentence styles reflect author preferences and are classified as periodic, loose, or minor. *Periodic* sentences begin with details and hold the main point for last. *Loose* sentences make their point early with initially placed subjects and predicates and follow with clarifying material. *Minor* sentences are incomplete sentences whose meaning is clear because preceding sentences provide explanatory material. A balance of sentence styles provides comfortable reading.

Effective sentences usually express a single idea and follow subject-verb-object order. Begin with the subject and its modifiers; follow with a verb, and add objects or complements to expand the meaning. The most direct subject-verb-object order occurs in the active voice when the subject initiates the activity.

The passive voice uses subjects as recipients rather than performers of actions. Although common in scientific writing, the passive voice is distracting because it requires extra words and can be difficult to translate. It is used to convey modesty by avoiding use of personal pronouns like *I* and *we*. This results in sentences like "*Accepted toxicological procedures were employed to analyze the specimens.*" However, "We conducted a toxicological analysis" is clearer and shorter. The convention of avoiding personal pronouns is changing and the use of *I* and *we* is becoming accepted in technical composition. Consequently, the verbose passive voice is losing favor.

Sentences are improved by searching for sentence faults during revision and editing. Look for sentences containing a double negative or lacking a subject or a predicate.

Double Negatives

Double negatives are statements with dual contradictory, disavowing, or denying words. They cause confusion. The major perpetrators are negative words like *no, never, not, none, nothing, hardly, scarcely,* and *barely* (Rozakis 2003), and contractions like *didn't, don't, haven't, doesn't,* and *couldn't* which camouflage the *not* word by deleting the letter *o.* The classic spoken double negative *"We don't have none"* is understood as *"We don't have any,"* but when literally translated, implies that *"We have some."* Double negatives are off limits in writing.

Fused Sentences and Sentence Fragments

Fused sentences present multiple jumbled thoughts and unrelated ideas. They can be indecipherable and are best corrected by reordering their points and rewriting them into several new sentences.

Sentence fragments are incomplete sentences or clauses that lack needed connecting words. Often they have the first word capitalized and end with a period, but they lack all the components of complete sentences. Some revision guidelines include:

- Read sentences aloud and rewrite them if you stumble or go back to figure out what is being said.
- Critique long sentences to see if they can be redesigned as two shorter sentences.
- Read sentences to identify their subjects, predicates, and objects, and determine if they are placed in that order.
- Identify sentences in the passive voice and try to rewrite them in the active voice.

After sentence faults are recognized and corrected, be sure the sentences are logically arranged in paragraphs.

Paragraphs

Paragraphs are topic-related sentence clusters that tell clear stories (Zeiger 1991). They signal topic transitions and offer readers momentary respites. They begin with a space or indentation and an opening introductory sentence which is followed by carefully chosen sentences with orderly and coherent details (Alred et al. 2003).

Effective paragraphs end with a summary that signals changing subject matter and prepares readers for a new topic. Extremely long paragraphs can stress readers. A variety of sentence lengths is appropriate within paragraphs. Paragraphs should:

- be limited to a single topic;
- open with an introductory sentence;
- conclude with a summary sentence that introduces the next paragraph; and
- vary in length and contain fewer than ten lines, or six long sentences.

Well-written paragraphs produce superb manuscripts.

Chapters

Chapters are major divisions of books. They usually contain introductory paragraphs, developmental paragraphs that detail the themes presented, short transitional paragraphs that signal topical changes, and concluding paragraphs with smooth transitions to material that follows (Steinmann and Keller 1999). Parallelism—one part of consistency—is vital in chapters. It is achieved by consistency in headings, lists, fonts, formats, tense, mood, and voice. Parallelism is discussed in chapters 1, 5, and 6. Book chapters are discussed in chapters 1 and 9.

Grammar Concluded

This chapter initially details some basic aspects of grammar and then goes into more detail than is essential for most writers.

CHAPTER 13

MEASURING WRITING PROGRESS

Introduction

Once you undertake writing improvement, you will make continual progress. Unlike schoolwork, music, or athletics, your mastery of writing won't be measurable by grades or applause. It will emerge gradually as growing confidence and professional advancement. You deserve feedback, encouragement, and incentive to intensify your quest. This can be accomplished by reviewing earlier writings and answering questions.

If you have already completed the writing scorecard at the end of this chapter, you have a baseline. If you didn't complete it when you read the introduction or chapter 2, do it now. Then review last year's writings.

An Annual Progress Review

An annual check will identify progress and areas needing work. Complete the scorecard. Then revise a paper, report, abstract, or chapter, and an already-sent letter or memo. Proofread the document and note what you've learned.

Look back, summarize your present status, and envision things to come. Call on the ghost of your writing future for a séance to brag about your progress. Also, outline a three-minute talk for use in casual conversation with unsuspecting colleagues. Resolve to move to greater heights and develop a new plan for further improvement.

The plan should address personal needs and aspirations, and commit to an unending pursuit of excellence. It will constantly change as you identify comfort zones and strategies that address your personal needs. The following activities will bolster a life-long pursuit of writing excellence.

Summarize Your Progress

Try writing a positive, personalized description of your developing writing style. It could sound something like this:

My efforts to achieve brevity, clarity, consistency, precision, accuracy, and flow have yielded an emerging style. My writing improves with revision and editing. I have reduced vagueness, ambiguity, and jargon, and replaced them with clear statements.

I work to eliminate wordiness and seek substitutes when the same word appears repeatedly. The substitutes are frequently better. I also replace many adjectives with more specific words and delete the words *very, crucial, many,* and *important.* I replace them with specific names, values, or explanations.

I am aware of the difference between precision and accuracy and have addressed both by replacing general terms with specific words and by conducting searches for accuracy of calculations, statements, and references. I work to see that each sentence flows smoothly from its subject to a nearby verb and then to a closely following object, and departs from subject-verb-object order only occasionally to break the monotony or to start sentences with an introductory phrase or clause.

My paragraphs begin with clear introductory sentences and close with summarizing sentences that introduce the next paragraph. I try to achieve a balance between long and short sentences and vary the length of paragraphs. Frequently, I introduce topics with sentences containing long strings of words separated by commas. Then using this as an outline, I follow with sentences detailing each point in that order. I tend to write long sentences. During revision I frequently replace the conjunction *and* with a period and begin a new sentence.

I am surprised at my progress. There are still areas needing improvement and I struggle to reduce the rambling and redundancy that mimics my speech patterns.

Once I begin a draft, each thought triggers new ideas and infinite possibilities unfold. There is a novel someplace in my mind and I have a folder for ideas about fictional characters. I eagerly anticipate new manuscripts and have five new project folders filled with notes. When they get overstuffed, I will replace them with three-hole notebooks.

I find proofreading intriguing and do it on airplanes and in waiting rooms and lobbies. No matter how many times I proofread a document, I still find double spaces, missing periods and commas, or misspelled words.

Since embarking on my writing crusade, I've found some previously unappreciated secrets that expedite my writing. They are:

- Miracles can be accomplished by drafting wildly without attention to detail.
- Many revisions are needed before my stuff is really good.
- I always knew more about good writing than I ever imagined.
- Project folders and notebooks are my life preservers.
- Dictionaries and thesauruses are my lifeguards.

Whenever I think about something that needs writing, I make notes about it in a folder and start collecting references. Once I decide on a title and message, my mind is bombarded with ideas that I jot down before forgetting them. After I draft a document, I lay it aside and keep dragging it out for revision. The harder I look, the more improvements I make.

I don't miss the TV ads and gossip sessions that used to occupy my time and energies. I devote that time to writing and do some every day. I experience occasional procrastination and relapsing writing blocks but overcome these by convincing myself that this effort will pay dividends.

I've gone far enough on the road to writing excellence that I've no desire to turn back. I know I still have lots of work to do. I will make these things happen because I believe in myself.

Now write **your** personal progress report. No fair copying mine. Set your report aside and don't peek till next year. Write a report each year and pat yourself on the back for continual progress. Remember, in order to appreciate and assess your progress, you must:

- proudly describe your evolving style periodically;
- save and number all major documents in notebooks;
- report annually to the ghost of your writing future; and
- list areas needing most improvement.

If you work these steps into a personal style you will make great progress. Each year, reassess your writing progress with the following scorecard.

Rate Your Writing Progress (a Scorecard)

This scorecard measures writing progress. It is best completed and saved early in the search for writing excellence, but whenever you first do it, a baseline for your journey to writing excellence will be established. Complete it annually to measure progress and encourage pursuit of writing excellence. While your assessments will be biased, they will surpass colleagues' comments in quantity and quality. You may copy this scorecard without copyright infringement.

For each of these criteria, honestly record a 1 to 10 score to estimate your proficiency someplace between disastrous (1) and perfect (10).

1. The Components of Writing Excellence
I can recite and briefly explain brevity, clarity, consistency, precision, accuracy, flow, and style and I have an idea of how to achieve them. I ignore these components during drafting but address them vigorously during revision, editing, and proofreading. My score is: _____.

2. Mental Preparation
I am pursuing writing excellence and have reprioritized my activities to include an hour of writing daily. I also joined a writers' group. My score is: _____.

3. Physical Preparation
I have secured a writing hideaway equipped with folders, notebooks, a dictionary and thesaurus, several writing manuals, and a computer. My score is: _____.

4. Drafting Wildly
I have realized that the best way to start is to draft as if I were the world's expert and not pause or make corrections until the first draft is complete. My score is: _____.

5. Multiple Revisions
Once drafts are completed I revise incessantly. This takes more time and effort than I imagined but it is the key to success. I have developed a revision procedure that I am comfortable with. Now I realize why some great writers revise

documents as many as 75 times. It is worth it. My score is: _____.

6. Editing
I am almost able to recognize when documents have been revised to death and it is time to edit. With practice and study, I'll soon be a good editor. My score is: _____.

7. Proofreading
I know what proofreading is all about. Since beginning line-by-line proofreading, I've found unique devices for checking consistency, accuracy, and precision. I am getting to be a pretty good proofreader but I'm still amazed at how many improvements in capitalization, punctuation, and spelling I achieve each time. My score is: _____.

8. Layaway Periods
I have begun to set aside documents after drafting, revising, and editing. I am continually amazed at the improvement opportunities that appear when I haven't looked at something for two or three days or more. After set-aside, I see so much that wasn't initially evident. I now start writing months before due dates to take advantage of this technique. My score is: _____.

9. Spelling
I find my spelling is improving by using simpler, more commonly understood words. I carry a list of my spelling challenges. The computer spell-checker is helpful but (like me), it is not perfect. Both of us need to be watching for typos and for correct selection of similarly sounding words with different meanings. My score is: _____.

10. Punctuation
I realize the emphasis and topic shifts I achieve verbally with pauses and volume changes can be mimicked in writing by careful punctuation. My new writing skills are helping me speak more effectively. I now realize writing requires different approaches than speaking. My score is: _____.

11. Grammar
I am familiar with the parts of speech and have some notions about their uses. I'm no grammarian and don't want to be, but I usually recognize when something I wrote is unclear and needs improvement. My score is: _____.

12. Meeting Deadlines
I have found that revision and editing greatly improve documents. This takes more time and effort than I imagined. I now begin writing well ahead of deadlines and I am still improving things as deadlines approach. I eventually have to say "This is it, next time I'll start sooner" and send it. My score is: _____.

13. Wordiness and Jargon
I now recognize my tendencies (shared by colleagues) to use extra words and write in terms that outsiders don't understand. I am working on this by cleansing my writings of meaningless modifiers, nonspecific adjectives and adverbs, and pretentious words. My score is: _____.

14. Different Kinds of Writing
I am aware of the subtle differences in the expectations and required levels of formality among the various writing mediums and use care to avoid email in situations requiring letters. I also am now conscious of the requirements for grant proposals, and the editorial requirements for scientific manuscripts and book chapters. My score is _____.

My total score is: _____.
I need to work hardest on items numbered _____ and
_____.
I will review my progress about a year from now, on
_____/_____/_____.

Conclusion
The maximum score on this assessment is 140. Repeat the evaluation process annually and you'll be amazed at your progress.

GLOSSARY
(WF/SBC indicates Writing Flaw / Should Be Corrected)

abbreviations. Shortened forms, usually the first few letters of a name, phrase, or title. Acronyms and initialisms are abbreviations. See also **acronyms**.

abridged dictionaries. Easily carried dictionaries with fewer definitions and less detail than encyclopedic or unabridged dictionaries.

absolute adjectives. Adjectives that stand alone without modifiers and cannot be compared with anything. Examples are: *unique, complete, pure,* and *unanimous.* See also **absolute words**.

absolute words. Words, such as *dead, prefect,* and *round,* that are complete and usually can't be modified or compared.

abstract nouns. Words that name intangible or nonspecific places or things. Contrast with **concrete nouns**.

abstracts. Concise summaries of documents or speeches.

accuracy. Correctness of data, processes, facts, names, addresses, dates, places, compounds, reagents, spelling, and arithmetic. Contrast with **precision**.

acronyms. Pronounceable words arising from the first letters of multiple-word names. Contrast with **initialisms,** which form non-pronounceable words.

active voice. Sentence arrangements in which the subjects perform the action. See also **voice** and **passive voice**.

adjectives. Words that describe, modify, or qualify nouns or pronouns. See also **articles, absolute adjectives**, and **proper adjectives**.

adjective sweep. A search of documents for nonessential adjectives.

adverbs. Words that describe, modify, or qualify verbs, adjectives, or other adverbs. Often formed by adding *ly* to adjectives, they describe the timing, frequency, and magnitude of actions and states of being.

agreement. Grammatical concordance among related words requiring subjects and verbs to coincide in person and number, and pronouns to match their antecedents in gender, number, and person.

ambiguity. Statements capable of being interpreted more than one way, often with opposite meanings.

ambiguous references. Word orders that leave uncertainty about which of two pronouns modify each of two nouns in a sentence.

ampersand (&). An abbreviated form that replaces the word *and* in names of corporations, professional organizations, websites, initialisms, and acronyms.

analogies. Similarities, resemblances, or comparisons with respect to sameness.

antecedents. Nouns, phrases, clauses, or sentences that are replaced by pronouns.

antonyms. Words with opposite meanings.

apostrophes('). Comma-like marks inserted after words to make them plural, to indicate possession, or to replace letters that have been omitted to form contractions.

appositives. Nouns or noun phrases that identify, clarify, or rename nouns directly preceding them in sentences and are often offset within commas, dashes, or parentheses, as in "Many tools, *that were needed,* were missing."

articles. The adjectives *a*, *an*, and *the* which precede nouns and are used to introduce them and focus their usage. See also **definite** and **indefinite articles**.

back matter. Also called end matter, material placed at the back of a book, monograph, or proposal and may include reference lists, bibliographies, glossaries, appendices, or notes.

biased language (WF/SBC). Word usage intended to convince readers of writer's views. It can include sexist, racist, ethnic, or other issues that may be offensive to readers.

biographic sketches. Brief statements of the background, education, professional experiences, and accomplishments of speakers, authors, or prospective employees. Compare with **curriculum vitae**.

boldface. Heavy, dark type used for headlines, headings, or emphasis.

braces { }. Punctuation marks used to indicate relationship between words or symbols in formulae or to achieve further separation of parenthetical or bracketed material.

brackets [ ]. Square parentheses used for insertions into constructions which are already in parentheses or for enclosing illogical material into sentences or inserting explanatory statements in material in quotations.

brevity. Briefness, conciseness, and shortness in statements.

bureaucratic language. Complex, difficult, and wordy language used in some governmental, academic, or business writings.

capitalization. The process of changing the first letter of a word to upper case (capital letter) to signify the beginning a sentence or begin formal names, titles, ranks, and academic degrees.

case. Modifications of the form of certain nouns, pronouns, or adjectives to indicate their relationship to the rest of a sentence. The subjective case (*he, she,* or *they*) is used as subjects of sentences; the objective case (*him, her,* or *them*) serves as objects of verbs or phrases; and the possessive case (*his, hers,* or *theirs*) indicates ownership.

clarity. Characteristics of clear statements that are void of vagueness or ambiguity.

clauses. Groups of related words containing both subjects and predicates that could be sentences if started with capital letters and ended with periods. See also **independent clauses** and **dependent clauses**.

clichés. Commonplace, trite, hackneyed, or vague expressions common in regional dialects.

colloquialisms. Informal words, expressions, or slang that is familiar in local or regional conversation but unfamiliar to others.

colons (:). Punctuation marks indicating that a pronouncement or quotation will follow, that something important is coming, or that the preceding statement will be clarified by what follows.

commas (,). Commonly used punctuation marks that signal pauses by separating introductory words, clauses, or phrases from sentences proper; by separating distinct parts of addresses; and by separating words in a series.

comma splices (WF/SBC). Also called comma faults, occur when two independent clauses are connected only by commas. If independent clauses are not made into sentences beginning with capital letters and ending with periods, they must be separated by semicolons or commas plus conjunctions like *and, but, or, nor,* or *for.*

common nouns. Nouns that name nonspecific classes of persons, places, or things. See also **proper nouns** and **concrete nouns**.

comparatives. Terms, usually adjectives or adverbs that indicate degrees of quality, quantity, or relationships. They appear in three forms: positives that qualify one thing or individual (good, fast, or accurate), comparatives that compare two things or individuals (better, faster, and more accurate), and superlatives that compare three or more things or individuals (best, fastest, and most accurate).

complements. Words, clauses, or phrases that explain, qualify, or expand the meaning of verbs or nouns.

complex sentences. Sentences that express at least two separate thoughts. See also **compound sentences** and **compound-complex sentences**.

compound adverbs. Multiple, often pretentious, joined words that appear as a single word. Examples include *heretofore*, *thereafter*, and *thereupon*.

compound-complex sentences. Lengthy sentences containing at least two separate thoughts expressed in at least two independent clauses and one dependent clause. See also **compound sentences** and **complex sentences**.

compound sentences. Sentences containing two or more independent clauses connected by appropriate punctuation. See also **comma splices**.

concise sentences. Sentences that present brief, clear, succinct statements.

concrete nouns. Nouns that name specific persons, places, or things. Contrast with **common nouns** and **abstract nouns**.

connecting adverbs. See **conjunctive adverbs**.

conjugations. Listings of the forms of verbs, usually presented in charts that show their correct use in each number, tense, voice, and mood.

conjunctions. Connecting words such as: *but, and, or, neither-nor*, and *either-or* that connect words, phrases, or clauses. See also **subordinating conjunctions** and **correlative conjunctions**.

conjunctive adverbs. Adverbs that connect independent clauses and thus serve as conjunctions. Examples are: *then, therefore, however, thus, accordingly, moreover*, and *besides*.

consistency. Harmony, uniformity, and compatibility in placement of words, sentences, paragraphs, and headings.

consonants. Letters other than the vowels *a, e, i, o,* and *u*. Consonants convey non-vowel sounds resembling *p, g, l, s,* and *r*. Compare these with **vowels** and **vowel sounds**.

constructions. The choice, arrangement, placement, and punctuation of words in sentences. See also **syntax**.

contractions. Words made shorter by replacing eliminated letters with apostrophes, as in *you are* contracted to *you're*.

coordinating conjunctions. Connecting words that join parts of sentences of similar function or equal rank.

copula. Verbs that connect subjects to predicates. See also **linking verbs**.

correlative conjunctions. Joining words that act in pairs to connect words or phrases within sentences as *either-or, neither-nor, not only-but also, both-and,* and *whether or not*.

curriculum vitae. A detailed description of professional careers, positions held, honors received, and lists of publications and speaking engagements. Compare with **biographic sketches**.

cut-and-paste errors. Mistakes that can occur in computer-generated documents during movement of text.

dangling clauses (WF/SBC). Clauses separated from the constructions they modify.

dangling modifiers. Adjectives, adverbs, or other modifiers separated from the constructions they modify.

dangling participles (WF/SBC). A participial phrase inappropriately attached to the wrong word or left unattached (dangling).

dangling phrases (WF/SBC). Phrases that are separated from the constructions they are intended to modify.

dangling sentences (WF/SBC). Sentences lacking an opening capital letter or a closing period or separated from concepts they are intended to clarify.

dashes (—). Horizontal lines used to signal sudden breaks or to segregate added explanations within sentences. Contrast with **hyphens**.

declarative sentences. Sentences that make statements.

definite article. The word *the*—the only definite article—a special adjective that indicates a single specific person, place, or thing. Contrast with **indefinite articles**.

demonstrative pronouns. Pronouns like *this, that, these,* and *those* that are often used without antecedents in speech. In scientific writing, they usually require identifiable antecedents even if the nouns they replace are in previous sentences.

dependent clauses. Word clusters with subjects and verbs but lacking complete meanings, and thus unable to stand alone because they lack modifiers needed to make sense. See also **independent clauses.**

determiners. Also called noun markers, perform grammatical activities by limiting nouns with respect to ownership (*mine, his, hers* and *its*) or quantity (*a, an,* and *the*). See also **articles**.

dialects. Distinct language characteristics of regions.

diction. Word choice. Diction determines the level of formality. It will vary depending on the message, the audience, the medium of transmission, and the tone chosen by writers' attitudes toward their readers.

direct objects. Nouns, pronouns, or their equivalents that receive the verb-described action generated by subjects of sentences or clauses. Contrast with **indirect objects**.

double-barreled hedges (WF/SBC). Use of two or more successive words suggesting doubt (such as *may possibly, possibly suspicious,* or *somewhat inconclusive*) to indicate uncertainty.

double comparatives (WF/SBC). Use of two comparatives, such as *more better* when one would suffice. See also **double superlatives**.

double negatives. Clauses or phrases that contain two negative words (such as *no, not, none, nothing,* and *nobody* or their contracted forms) so as to effectively cancel one another and produce a meaning opposite from that intended.

doublespeak. Purposely vague, confusing, or evasive language.

double superlatives (WF/SBC). Use of two superlatives, such as *most best,* when one would suffice. See also **double comparatives**.

draft. The first rough copy of a document. See also **drafting**.

drafting. The process of putting thoughts into sentences, however disjointed, on paper, as the initial step in document preparation. See also **draft**.

editing. The process of reviewing documents to make improvements or achieve compliance with publishers' standards.

ellipses / ellipsis dots (…). Marks inserted in text to indicate thoughtful pauses, intentional omission of words, or marked shifts in topics.

enclosure notice. A statement naming additional material included with a letter, placed below the signature and preceded by the designation *Enclosure*:

errata. Lists of corrections published in subsequent issues of journals or inserted in books.

et al. A Latin abbreviation meaning *and other people* used after the names of first authors in citing multiple-authored references.

euphemisms. Modifications of harsh, offensive, or socially unacceptable terms or figures of speech in order to sidestep reality or soften undesirable impressions.

exclamation points (!). Punctuation marks inserted to gain attention, issue commands, or indicate urgency.

exclamatory sentences. Sentences or one-word insertions that express urgency, strong feeling, or commands and usually punctuated with exclamation points.

figures of speech. Expressions such as analogies, euphemisms, metaphors, or similes that use words in an atypical way to convey nonliteral meanings, often to make a point.

flow. Smooth, liquid-like transition from word to word, sentence to sentence, and paragraph to paragraph, that results when carefully chosen words are placed in logical order.

folio. A page number, the number of pages in a book.

fonts. Assortments of sizes and styles of type available on computers.

foreword. A preliminary statement or introduction to a book usually written by someone other than the author. Compare with **preface**.

format. Organization, arrangement, and physical appearance of material in writings, including print style (font), letter style, section headings, and margins.

fragments (WF/SBC). Incomplete sentences or clauses that begin with capital letters and end with a punctuation mark but lack needed connecting words.

freelance writers. Writers that sell their work.

FTP. File transfer protocol, an electronic device for transferring files via the Internet.

fused sentences (WF/SBC). Indecipherable, often non-repairable sentences presenting multiple unrelated thoughts and excess ideas, also called **run-on sentences**.

gender. Designation of nouns, pronouns, and adjectives as masculine, feminine, or neuter.

gender bias. Use of the word *he* to replace nouns like *God* that could also be feminine.

gerunds. Nouns or noun forms used to name persons, places, or things and possessing some characteristics of both nouns and verbs, formed by adding *ing* to verbs.

gerund phrases. Phrases that use gerunds (*ing verbs*) as subjects or direct objects.

glossary. A list of definitions or explanations of words in the context used in a document.

gobbledygook (WF/SBC). Stuffy, vague, complex, and pretentious writings laden with jargon, buzzwords, legalese, bureaucratese, and multiple modifiers in an effort to sound scientific, official, or authoritative.

grammar. Rules for placement of words in sentences for function, position, and meaning.

grammatical gender. See **gender**.

headings. Titles, headlines, or subtitles placed above paragraphs, sections, chapters, or news articles, and usually in boldface.

homonyms. Easily confused words such as *weather* and *whether* that sound alike but have different meanings and are often spelled differently. Contrast with **synonyms**.

hyphens (-). Small single horizontal marks used to separate parts of compound words or syllables of divided words. Contrast with **dashes**.

idioms. Peculiar word collections with nonliteral meanings that are readily understood by speakers of regional dialects.

imperative mood. Verb usage that conveys writers' intentions to present commands, directions, or requests. See also **mood**.

imperative sentences. Sentences that issue directions, orders, or commands and sometimes end in exclamation marks. Subjects of imperative sentences are often omitted and understood to be *you*.

indefinite articles. The singular adjectives *a* and *an*. They indicate oneness (singular number) and usually modify nonspecific nouns. Compare with **definite article**. See also **determiners**.

indefinite pronouns. Pronouns that refer to unnamed persons (such as *anyone* or *no one*), places (as *anyplace, someplace,* or *anywhere*), or things (such as *anything*).

independent clauses. Clusters of words containing appropriately ordered subjects and predicates so they could be sentences if punctuated with capitals and periods.

indicative mood. Verb usage that reflects writers' wishes to state facts or convey questions. See also **mood**.

indirect objectives. Nouns or noun forms that are receivers of the action of transitive verbs and state to *whom* or to *what* the direct objects were passed.

infinitives. Verb forms, usually accompanied by the word *to*, that are neutral regarding tense, number, and person, and often perform as nouns in sentences, sometimes called *verbal nouns* because they can behave as verbs or nouns.

infinitive phrases. Phrases that are introduced by infinitives, and often use infinitives as their subjects. For example: "*To intervene* surgically could be catastrophic." See also **infinitives**.

initialisms. Shortened forms of multi-word names using the first initial of each component to form a non-pronounceable cluster of letters which must be pronounced individually, as in FDA. When pronounceable as a word, they are called acronyms.

interjections. Words that are inserted, often without a logical grammatical basis, into sentences to indicate emphases, emotions, or exclamations.

interrogative pronouns. Pronouns used in asking questions. Examples are: *who, whom, which,* and *what.* See also **interrogative sentences**.

interrogative sentences. Sentences that ask questions and end in question marks.

intransitive verbs. Verbs that completely describe actions without requiring objects.

inversions. Sentence structures that deviate from subject-verb-object format and are best avoided but used occasionally to achieve emphasis or variety.

italics. A typestyle that uses *vertically slanted* letters to emphasize words, titles, figures, or explanatory phrases.

jargon. Specialized technical terminology used for communication within groups and professions.

layaway periods. The essential, but often neglected, writing strategy that involves setting documents aside for a period before reworking them.

linguistic science. The science of language structure and makeup that focuses on speech and dialect.

linking verbs. Also called copula verbs, such as *appear, become, smell,* and *taste,* that connect parts of sentences by describing occurrences or states of being. They are usually followed by nouns, pronouns, or adjectives.

loose sentences. Sentences that make their point initially and continue on with details, attached modifiers, or qualifications.

main clauses. Independent clauses that could stand alone as sentences if properly capitalized and punctuated with a period.

malapropisms. Ridiculous, absurd, or ludicrous misuse of similar sounding words with distinctly different meanings in an effort to be humorous.

metaphors. Figures of speech that suggest similarities or relationships between dissimilar things, ideas, or words. They are usually used to make a point. The well-known item helps readers understand the less familiar one. See also **mixed metaphors**.

minor sentences. Another name for incomplete sentences.

misplaced modifiers (WF/SBC). Words, phrases, or clauses that modify the wrong word.

mixed metaphors. Strings of metaphors or comparisons that are inconsistent or incongruous.

modifiers. Words, phrases, or clauses that define, describe, limit, or qualify the meaning of words or other passages. See also **misplaced modifiers**.

mood. Grammatically variant forms of verbs that indicate writers' feelings or attitudes. Mood shapes sentences to carry implied facts or questions (indicative mood); commands or directions (imperative mood); and wishes or desires (subjunctive mood).

nominalization. The process of converting non-noun words to a noun structure, as *vaccinate* (verb) to *vaccination* (noun) or *cultivate* (verb) to *cultivation* (noun). See also **nominals**.

nominals. The grammatical name for non-noun words, phrases, or clauses serving as subjects and thus used in the nominative case.

nouns. Words that name persons, places, things, or ideas.

noun clauses. Clauses that function as nouns.

noun phrases. Phrases that contain nouns and can serve as subjects or objects.

number. The grammatical quality of nouns, pronouns, or verbs that designates whether the persons, places, or things discussed are singular or plural.

objective case. Sometimes called the accusative case, the forms nouns take when serving as objects of verbs or prepositions.

objects. Nouns, noun forms, or pronouns that complete the meaning of transitive verbs or prepositional phrases. See also **direct objects** and **indirect objects**.

orthography. Spelling.

parallel structure. See **parallelism**.

parallelism. A form of consistency, also called parallel structure, that enhances flow by assuring that sentence elements have common form (tense, number, voice, and mood) particularly when words, phrases, clauses, or sentences are used in series, lists, or tables of contents.

parentheses ( ). Curved vertical marks used to set off parenthetical expressions. See **parenthetical expressions.**

parenthetical expressions. Explanatory or amplifying words, statements, or digressions placed within parentheses to expand on the meaning of passages usually not critical to understanding.

participial phrases. Participles, along with modifiers and objects that function as adjectives, as in: "The resident, *serving under the laboratory director*, conducted the tests." See also **participles**.

participles. Words (usually verb forms) that function as modifiers, such as the *running* horse. Present participles indicate something is currently going on and end in the suffix *ing*; past participles indicate something has already happened and usually end in *ed*, and perfect participles indicate actions that have previously happened but are continuing, or ongoing, or planned for the future. See also **dangling participles** and **participial phrases**.

parts of speech. The classes to which words are assigned on the basis of function, form, meaning, and grammatical function. Many words can serve in several parts of speech.

passages. Small sections of sentences, paragraphs, clauses, or phrases that are singled out for attention.

passive voice. Verb usage indicating that the subject is the recipient of the action. See also **voice**. Compare with **active voice**.

PDF. An initialism or abbreviation for *portable document format*, often written in lower case letters as pdf. It is a computer-generated file that adheres to a standard format so it can be forwarded as email with no loss of words or pictures.

pedagogy. The art and science of teaching.

periodic sentences. Sentences that present subordinate details at the beginning and save the main points until the end.

periods. The dots used to end sentences and provide a written counterpart of the pause or full stop that speakers use when new thoughts are introduced.

person. The quality of pronouns that indicates their relationship to readers. The writer represents the first person (*I*); the reader is the second person (*you*); the third person is the individual or thing written about (*he, she,* or *it*).

personal pronouns. Pronouns that replace nouns that name people or things. Examples are: *I, me, you, he, she, it, we,* and *they*.

phrase. A group of related words that has a subject or a predicate but not both. See also **noun phrases**, **verb phrases**, **prepositional phrases**, **participial phrases**, **infinitive phrases**, and **gerund phrases**.

plurals. Words implying or containing more than one item. Contrast with **singular**.

possessive case. Nouns or noun forms that indicate possession (ownership) and similar relationships, such as a manufacturer (Ford's cars), a writer (Clinton's book), or a record holder (Bannister's record).

possessive pronouns. Pronouns that indicate ownership. Examples are: *my, mine, your, yours,* and *his* and *hers*. See also **pronouns** and **personal pronouns**.

PowerPoint presentations. Talk outlines, often in bullet format, prepared using computer programs and capable of transfer to projectors for creation of screen images that can include graphics.

precision. Scrupulous detail, exactness, correctness, certainty, and definiteness attained by using specific names and measurements rather than general terms. Contrast with **accuracy**.

predicate nominatives. Also called predicate nouns, are nouns or pronouns connected to subjects by linking verbs so as to rename or provide additional information about them.

predicates. The operative words of sentences, verbs, or verb forms that express action or being, and are the portions of sentences containing the main verb and its modifiers.

preface. Introductory remarks, usually written by authors, that explain the basis, organization, and purpose of books. See also **foreword**.

prefixes. Letters placed in front of root words to modify or change their meaning. Common prefixes are *anti*, *macro*, *mega*, *pre*, and *pro*.

prepositional phrases. Modifying phrases consisting of prepositions and their objects (nouns or pronouns and their modifiers). Prepositional phrases can act as adjectives or adverbs. See also **prepositions**.

prepositions. Linking words such as *at*, *by*, *for*, *in*, and *to*. They illustrate relationships by connecting nouns or pronouns to other words.

press releases. Candid, transparent written information of public interest that is given to the media for their use. They are often ignored or edited.

pretentious words. Usually long, self-promoting, and flattery-seeking words.

print-on-demand (POD) publishers. Self-publishing companies able to print any number of books to fill orders quickly. See also **publication on demand** and **vanity presses**.

pronouns. Words that take the place of nouns. See also **demonstrative pronouns**, **indefinite pronouns**, **personal pronouns**, and **possessive pronouns**.

proofreaders' marks. Signs placed in margins to identify required corrections. Lists of proofreaders' marks appear in grammar books and dictionaries.

proofreading. A final, detailed search, conducted by editors, to identify errors before documents are printed and increasingly done by authors before submission.

proofs. Written materials that are ready for printing, often called page proofs or galley proofs.

proper adjectives. Adjectives derived from proper nouns. Examples are: *European* exports, *Congressional* parties, and *Harvard* graduates. See also **proper nouns**.

proper nouns. Words such as President Lincoln, New York City, or the Cowboy Museum, that name specific persons, places, or things.

proposal. Detailed descriptions of writings and their author(s) presented to publishers to convince them to consider publishing a manuscript. Compare with **query letters**.

publications on demand. Production of books at the author's expense, usually places marketing responsibility upon the writer. See also **vanity presses** and **print-on-demand (POD) publishers**.

publishers' styles. Formats, headings, fonts, and other design details required by publishers of journals and books.

punctuation. Use of marks such as periods, commas, colons, and semicolons to direct the speed and intent of sentences by signaling pauses and indicating changes in direction.

query letters. Brief letters describing books to determine the interest of publishers. Compare with **proposals**.

question marks (?). Punctuation marks used to follow direct questions or suggest doubt.

quotation marks ("..."). Punctuations used to indicate another person's words are used.

rambling sentences (WF/SBC). Exceedingly long and cumbersome sentences that contain more information than can be comfortably handled.

redundancy. Nonessential repetition of words, sentences, or information.

reflexive pronouns. Pronouns ending in the words *self* or *selves* (*myself, yourself,* and *themselves*) that refer back to noun subjects, indicating that the noun acts upon itself, as in "I hurt *myself*."

relative pronouns. Pronouns that take the place of nouns or substitute for nouns and serve to relate one part of a sentence to another by linking dependent clauses to main clauses. They include: *who, whom, whose, which, what,* and *that*.

revision. The systematic rearrangement and rewriting of drafts to achieve brevity, clarity, consistency, accuracy, precision, and flow.

root words. Word segments with specific meaning, many of Greek or Latin origin, that combine with prefixes or suffixes to form words.

run-on sentences (WF/SBC). Two or more complete ideas that are poorly joined. See also **fused sentences** and **comma splices**.

semantics. The study of word meanings and arrangements.

semicolons (;). Punctuation marks suggesting word separation greater than that attained by commas and also used to separate series of words, phrases, or clauses which themselves contain commas or other punctuations.

sentence faults (WF/SBC). Sentences that lack a subject or a verb and confuse readers. See also **sentence fragments, run-on sentences,** and **dangling modifiers**.

sentence fragments (WF/SBC). Grammatically incomplete sentences punctuated and used as complete sentences.

sentences. The principal ingredients of writing; orderliness and flow are key indicators of quality. They use subjects and predicates to express complete thoughts.

sic. A Latin term meaning *so, thus,* or *intentionally written. Sic* is usually placed in brackets or parentheses to indicate the writer recognizes a misspelling or erroneous statement but is quoting it as it originally appeared.

similes. Figures of speech that compare two dissimilar things using the words *like* or *as.* See also **metaphors**.

simple sentences. Sentences that include one independent clause and express a single thought.

singular. Word forms that denote only one person, thing, or instance.

slashes (/). Diagonal lines or slash marks that indicate omitted words or letters (feet/second); separate numerators from denominators (1/2); or separate parts of dates (6/28/05).

spam. Unsolicited email messages or electronic junk mail, also used as a verb meaning to send emails to multiple unsuspecting recipients.

speech recognition programs. Computer programs which, after setup exercises, recognize and translate speakers' words into typewritten documents at speeds exceeding typists'.

spell-checkers. Computer programs that identify and improve incorrect and misspelled words.

stet. Proofreaders' mark meaning "save the existing text" that identifies erroneous corrections which should stand as originally written.

structure. The organization, arrangement, pattern, and makeup of writings.

styles. Individual communicative methods that set writers apart. See also **publishers' styles**.

subject. The noun or pronoun serving as the key element of a sentence by naming the person or thing performing the action or described by the main verb.

subject-verb agreement. Required to be matching within sentences of subjects and verbs with respect to number (singular or plural) and person (first, second, or third).

subject-verb-object order. Sentence arrangement involving an opening subject followed closely by verb forms and then by objects and their modifiers.

subjunctive mood. Usage of verb forms that suggest uncertainty, doubt, wish, possibility, contingency, or writers' beliefs that something is contrary to fact. See also **mood**.

submission. Presenting writings for review or consideration.

subordinate clauses. Clauses that modify or expand main clauses but nonetheless depend upon them for meaning. Usually they connect clauses of differing levels of importance. See also **dependent clauses**.

subordinating conjunctions. Words that connect sentence elements of different weights to make one less important than the other. Examples are: *when, where, before, as, since*, and *because.*

succinctness. A quality of compact and precise writing characterized by brevity.

suffixes. Letters added behind root words to change or modify their use or meaning. Examples are: *ese, like, ly, ous*, and *wise*.

synonyms. Words with similar or identical meanings, often sought in a thesaurus, to avoid word repetition.

syntax. The selection, arrangement, and relationships of words in the formation of phrases, clauses, and sentences.

tense. Verb timing indicating the past, present, or future (simple tenses) and past perfect, present perfect, or future perfect (perfect tenses).

transitions. The techniques for creating a smooth flow from concept to concept between words, sentences, and paragraphs, achieved by choosing connecting words (such as *however, moreover, furthermore,* and *in addition to*) or phrases to show relationships between ideas.

transitive verbs. Verbs requiring a direct object to complete their meaning.

typeface. All type of a single design. See also **typestyles**.

typestyles. Bold, italic, or underlined letters or words available through icons on computer toolbars.

typographical errors (WF/SBC). Writing errors (also called typos) resulting from typing or keyboard slip-ups that produce misspellings or spacing oversights, often detected during proofreading.

unabridged dictionaries. Extensive encyclopedic dictionaries that contain more words, details, and acronyms than abridged versions. They are usually too heavy to be easily carried.

upper case. Capital letters or caps.

URL. Uniform resource locator; the specific computer addresses of websites.

vanity presses. A gradually vanishing name for self-publishing. See also **publication on demand**.

verbals. Words derived from verbs that mimic or act as nouns, adjectives, or adverbs. They include gerunds formed by adding *ing* to verbs, infinitives, and participles.

verb phrases. Phrases that function as verbs or verb forms.

voice. Verb usage and placement that designates subjects of sentences as doers (active voice) or as receivers of the actions or descriptions (passive voice).

vowels. The letters *a, e, i, o,* and *u.* So named because in English they are pronounced so as to produce vowel sounds. See **vowel sounds**.

vowel sounds. Speech sounds resulting from direct passage of air through the open mouth and most commonly produced by common pronunciation of the vowels *a, e, i, o,* and *u.*

WF/SBC. Initialism for Writing Flaw / Should Be Corrected.

who/whom dilemma. Confusion as to whether to choose the word *who* or *whom. Who* is used in nominative settings and the word *whom* is used in the objective.

wordiness. Use of excessive or pretentious words.

word placement. The order of words and phrases in sentences. Excellent word placement requires minimizing the distance between subjects and verbs, verbs and their objects, and modifiers and the words they modify. See also **syntax**.

writer's clout. Professional enthusiasm and spirit generated by writing skill and confidence that triggers improved public speaking, expanded knowledge, and increased workplace influence.

writing blocks. Readily curable inability to write, caused by temporary brain paralysis and frustration. Most writers experience occasional writing blocks. Compare with **writer's clout**.

* * *

NOTES

BIBLIOGRAPHY

Agnes, M. 2002. *Webster's New World Dictionary and Thesaurus.* 2nd Edition. New York: Hungry Minds Inc.

Alred, G. J., Brusaw, C. T., and Oliu, W. E. 2003. *The Handbook of Technical Writing.* 7th Edition. New York: St. Martins Press.

Alward, E. C. and Alward, J. A. 2000. *Punctuation; Plain and Simple.* New York: Barnes & Noble.

Andersen, R. 2001. *Powerful Writing Skills.* New York: Barnes & Noble.

Austin, T., editor. 1999. *The Times Guide to English Style and Usage.* London: Times Books, HarperCollins.

Beckett-McWalter, J. et al. 2002. *Vocabulary and Spelling Success in 20 Minutes a Day.* 3rd Edition. New York: Learning Express.

Carnegie, D. 1998. *Dale Carnegie's Lifetime Plan for Success. How to Win Friends and Influence People* and *How to Stop Worrying and Start Living.* New York: Galahad Books.

Cazort, D. 2002. *Under the Grammar Hammer.* 2nd Edition. New York: Barnes & Noble.

Council of Scientific Editors. 2006. *Scientific Style and Format. The CSE Manual for Authors, Editors, and Publishers.* 7th Edition. Reston, VA: Council of Scientific Editors.

DeVries, M. A. 2002. *The Practical Writer's Guide.* New York: Barnes & Noble.

Elliott, R. 1997. *Painless Grammar.* Hauppauge, NY: Barron's Educational Series.

Hjortshoj, K. 2001. *Understanding Writing Blocks.* New York: Cambridge University Press.

Hodges, J. C. et al. 1998. *Harbrace College Handbook.* 13th Edition. Fort Worth, TX: Harcourt Brace College Publishers.

Hopper, V. F. et al. 1990. *Essentials of English.* 4th Edition. Hauppauge, NY: Barron's Educational Series.

Infinity Publishing.com. 2007. *Become A Published Author.* West Conshohocken, PA: Infinity Publishing Company.

Kane, T. S. 2002. *The Oxford Essential Guide to Writing.* New York: Berkley Books.

Maas, G. S. (Editor). 2000. *Random House Webster's English Language Desk Reference*. 2nd Edition. New York: Random House Inc.

McCollister, J. 1999. *Writing for Dollars*. New York: Barnes & Noble.

Merriam-Webster Incorporated. 2001. *Webster's Business Writing Basics*. Springfield, MA: Federal Street Press, Division of Merriam-Webster.

Meyer, H. E. and Meyer, J. M. 1993. *How to Write: Communicating Ideas and Information*. New York: Barnes & Noble.

O'Hayre, J. 1966. *Gobbledygook Has Gotta Go*. Washington: U.S. Government Printing Office. (Document O-206-141).

Pfaffenberger, B. 2003. *Webster's New World Computer Dictionary*. Indianapolis, IN: Wiley Publishing Inc.

Random House Publishing Company. 2003. *Random House Webster's Handy Grammar, Usage & Punctuation*. 2nd Edition. New York: Random House Reference.

Rozakis, L. E. 2003. *The Complete Idiot's Guide to Grammar and Style*. 2nd Edition. New York: Alpha Books.

Sabin, W. A. 2001. *The Gregg Reference Manual*. 9th Edition. New York: Glencoe McGraw-Hill.

Shertzer, M. 2001. *The Elements of Grammar*. New York: Barnes & Noble.

Shimberg, E. F. 1999. *Write Where You Live: Successful Freelancing at Home*. Cincinnati, OH: Writer's Digest Books.

Steinmann, M. and Keller, M. 1999. *Good Grammar Made Easy*. New York: Random House Value Publishing Inc.

Strunk, W. and White, E. B. 2000. *The Elements of Style*. 4th Edition. New York: Longman.

Thurman, S. 2003. *The Only Grammar Book You'll Ever Need*. Avon, MA: Adams Media Corporation.

University of Chicago Press. 2003. *The Chicago Manual of Style*. 15th Edition. Chicago: University of Chicago Press.

Weverka, P. 2001. *Word 2002 for Dummies*. New York: Hungry Minds Inc.

Woods, G. 2002. *Research Papers for Dummies*. New York: Hungry Minds Inc.

Zeiger, M. 1991. *Essentials of Writing Biomedical Research Papers*. New York: Health Professions Division, McGraw-Hill.